Webplicity 2.0

The Critical Guide to Successful Web Strategies

Bill Young

iUniverse, Inc.
New York   Bloomington

Webplicity 2.0

The Critical Guide to Successful Web Strategies

iUniverse books may be ordered through booksellers or by contacting:

iUniverse
1663 Liberty Drive
Bloomington, IN 47403
www.iuniverse.com
1-800-Authors (1-800-288-4677)

ISBN: 978-0-595-53289-6 (pbk)
ISBN: 978-0-595-63344-9 (ebk)

Printed in the United States of America

iUniverse rev. date: 11/17/08

Contents

1.0 Analysis
    Case Study: Water Utility Client
    1.1 Past Experience
    1.2 Your Purpose
    1.3 Real Goals
    1.4 Personnel Involved
    1.5 Define Success
    1.6 Establish ROI
    1.7 Choosing A Firm
    1.8 Strategy Meetings
    1.9 Creating a Creative or Technology Brief

HOW TO USE THIS BOOK

Webplicity 2.0 dives into more details related to web applications, both their effect on online marketing initiatives and operational efficiencies. The first book shed some light on the subject of how to build a Web project, be it a Web site or complex Web application. We showed the good, the bad, and the ugly when it comes to Web sites and Web applications. During the past two years, we've seen the explosion of social media and the impact on marketing and business objectives. The Web continues to be filled with examples of ineffective work. Now the challenges not only include poor design and bad programming but add in inept integration of online marketing tools with business objectives. Firms are still trying the "throw it on the wall and see what sticks" tactical plan. Although our first book helped us take our knowledge and process and save the world, version 2.0 might save the universe. Well, maybe not the entire universe, but the virtual world. Our main goal still continues to be an effort to save business executives, owners, and front line marketing and IT personnel from nightmare Web projects.

Developing a new Web project is still a tricky process no matter how you go about it. Some would think that the Web is maturing, but on the contrary it is becoming more diverse and complex every day. According to the firm Technorati, there are over 70 million weblogs and 120,000 new ones are being created worldwide each day. That's about 1.4 blogs created every second of every day. Maturity? We'd describe it as crazy.

XploreNet team has been implementing Web-based solutions since 1997 and the process is still a difficult one. Version 2.0 updates the improvements to our process and recent experience and success in implementing unique new solutions that incorporate both web applications and online marketing tools (tied of course to social media issues). Furthermore, we reveal new *secrets* and present you with many ideas. We still won't reveal to you the "secret sauce," that continues to be purely proprietary, and as we've seen our competitors still don't "get it". But in version 2.0, we'll offer you valuable insight that may very well make the difference in your Web project being a success or a failure. The key to getting started on a successful Web project or properly integrating online marketing tools is a well-developed Web plan. What to include in it is the hard part.

In this book, we take you through the process of developing a Web plan step by step, and show you how your Web strategies, business goals, and specific projects fit into the plan. Version 2.0 is focused more on reducing internal operational costs and the drive part of our process (identifying and increasing users). Whether you are the CEO, Business Owner, VP/Director of Marketing, CIO, CTO, Director of IT, VP of Communications, Board member, or Executive Director, version 2.0 can help you complete your most important business goals. Furthermore, whether you are a small firm attempting to redesign a basic Web site or a large organization implementing a complex tool, the five modules (if completed properly) will ensure the most complex Web project runs smoothly.

These five modules will help you through the process:

MODULE 1:	Covers the Analysis Phase of the process, showing you how to get started on the right foot.
MODULE 2:	Contains the Blueprint to success. Just as an architect completes blueprint drawings for a dwelling, you must complete both a site and project blueprint in order for your Web project to be successful.
MODULE 3:	Covers the Construction Phase – no need to get your hands dirty or buy work gloves.
MODULE 4:	Discusses the Promotion Phase of any new, Web-based tool.
MODULE 5:	Covers the steps to putting everything together and provides an example of a successful Web plan outline.

The outlined processes in this book are tried and true methods that help facilitate success for any sized firm, on any sized project. Almost 11 years of experience tell us that we "get it".

The best way to use this book is to understand the entire Web plan format and use it on your next Web project.

In the end, our hope is that this book helps you put all the pieces of the puzzle together to successfully launch your next generation Web project and complete the proper web strategies.

Welcome to the next step in the Web.

Visit www.webplicitybook.com click on to download
the full web plan **enter the code web20**

ABOUT THE AUTHORS

Bill Young has assisted hundreds of firms since 1996 with the development and implementation of successful Web strategies.

The modules in this book came about through Young's experience, as Chairman and Co-founder of XploreNet (www.xplorenet.com), and through trial and error. XploreNet is a leading provider of technology and marketing integration services in the Rocky Mountain Region, serving customers like American Water Works Association Research Foundation (AwwaRF), Pak Mail Centers of America, and GE.

After graduating from Colorado State University with a Bachelors degree in Business Management, Young co-founded XploreNet in 1997 with $700, a few credit cards, and a dream. Young's vision was to build a consulting company that brings a human perspective to a very technology-driven industry.

Young was honored by the Denver Business Journal as a winner of the "Forty Under 40" award in 2000. In addition, he co-authored the book, <u>Brick and Mortar, Click and Order - The Encyclopedia of Retail and Ecommerce</u>™, teaming with Keven Bernard (1995 Retailer of the Year). Their idea was to bring the worlds of traditional retailing and e-commerce together in a dynamic hybrid approach to business.

Co-Author Dave Dixon is a member of the XploreNet management team and serves as the COO and VP of Client Services.

Dixon is no stranger to the world of marketing and advertising having spent the first part of his career with the worldwide advertising agency, Young & Rubicam, where he was ultimately responsible for managing multi-million dollar regional advertising and marketing efforts for a national client. He then moved onto AT&T Broadband (now Comcast), who at the time was the nation's largest cable operator, where he managed the development and execution of numerous national marketing campaigns for their cable products. Most recently, before joining XploreNet he was a member of the product development team for a major health insurance carrier.

Dixon has a Bachelors Degree in Business from Miami University and has an MBA from the University of Denver, Daniels College of Business.

Co-Author Michael Sevilla is founder of Gravity Metrics (which was purchased by XploreNet in 2004).

Sevilla serves as VP of XploreNet and under his direction, Gravity Metrics was launched with the sole purpose of demonstrating methods to increase sales and achieve the best possible results from a Web project and all online marketing initiatives. Many of GM's past clients come from the advertising, e-commerce, hospitality/resort, and online training industries.

Michael has an extensive background in marketing, brand management, and product management/development for the consumer goods and technology industries, and received an MBA for International Marketing and Finance from Thunderbird.

Visit xplorenet.com or call 720.221.9214 for more helpful ideas.

WEB PLAN OVERVIEW

HOW DO YOU BUILD A WEB SITE OR ONLINE TOOL? At XploreNet we've heard this question countless times. A better question might be how do you build a *successful* Web site? As you probably experienced, going about it the wrong way is more commonly found.

Why do we get this question so often? The reality is that most firms are still using the trial and error process for web projects. There is enough information available on the Internet to fill Coors Field five times over. Yet figuring out the best solutions and paths to obtainment can give anyone a headache.

What is a Successful Web Project?

Why do only half of all Web projects fall under the definition of a successful project? According to CIO Insight, only 40 percent of CIOs say they have a Web project plan in place. No wonder only two-thirds of projects are likely to come in on budget and less than 60 percent are likely to achieve their ROI targets.

In version 2.0, we're attempting to answer this question by providing a map to help guide your Web project from the **present** to the **future**. We will help you connect your web presence to your online marketing strategies. Most successful businesses have a business plan and some even have marketing plans, but very few have a real Web plan. In order to reach the ultimate goals for your organization, you must have a business plan, marketing plan, and a Web plan.

How?

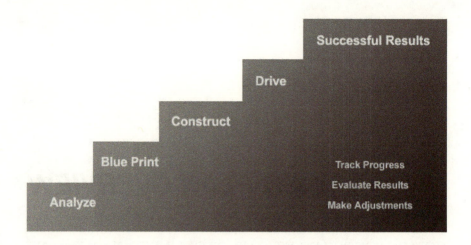

We'll show you. We cover five modules in this book:

- ▸▸ Module 1: Analysis
- ▸▸ Module 2: Blue Print
- ▸▸ Module 3: Construct
- ▸▸ Module 4: Drive
- ▸▸ Module 5: Putting it all together

It may seem like a daunting process to complete a detailed plan; however, the success of your new Web project may depend on your fully developed Web plan.

"This is the first time our investors have had online access to information about their accounts."We feel it's a valuable tool that will enable them to maintain more control over their investments. "

Director of Marketing
Stacy Dysart

MODULE 1

ANALYSIS

1.0 Analysis

This module discusses the first phase of understanding how your Web project fits into your Web strategies and defines your Web plan. As you read, you will quickly understand whether your Web strategies are on life support or are dead. In today's competitive business environment, companies that successfully implement a new redesign or build a complex Web tool, routinely look back to the initial Analysis Phase as the foundation to their success. A well-defined Web strategy includes an interactive, customer-focused Web presence, "user-friendly" Web applications, and proper interaction of online marketing tools. If you overlook something in the Analysis Phase, then the rest might be meaningless.

Figure 1.0

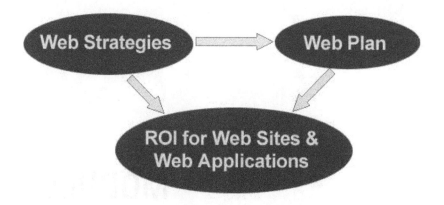

"We greatly appreciate the relationship between the Foundation and XploreNet."

Webmaster AwwaRF
Sue White

Business Web sites today need a plan in order to remain competitive. Early Web sites served simply as glorified online brochures. Now they must produce something tangible. Of course the site must be visually appealing, but the content must also have a point.

This means: multiple portals for different audiences; secure, encrypted e-commerce sites; secure back-end systems for storing employee, vendor, and customer information; real-time data integration to inventory databases, and customer resource management (CRM) systems; and simple processes for usability. Issues regarding design and complexity change based on the size and scope of your firm; however, everyone wants an interactive and successful Web project. Now there exists a new area called online marketing, which includes many different tools. We refer to it as online marketing, but the current buzzword is "social media". Firms must understand and properly integrate online marketing tools to fit their objectives. For instance, we see firm after firm throwing money and effort at one of these tools called, Social Media just to look like they are cutting edge. Examples include MySpace pages, blogs, podcasts, and networking sites (to name just a few). We'll cover more on these tools in the Drive phase. However, if you do not ask the proper questions during the analysis, you'll end up with tools that don't connect your customers to your brand nor do they help facilitate revenue growth or more customers.

The XploreNet Web Plan was developed during the past ten years. It is based on our development processed called ABCD's of development. This plan incorporates a detailed Web strategy that covers how to build the correct Web presence, integrate the proper applications, and promote the site to the right audience, <u>for maximum results</u>. No matter what a business or organization offers, it must have the right strategies and a professional Web presence in order to be competitive in today's business world. It is critical for profit-oriented, traditional corporations that sell products to have an outstanding presence, but non-profits and service-providing organizations, such as lawyers and charities, must have a quality Web presence as well.

The Analysis Phase is critical to marketing vice presidents and directors who must understand how each marketing initiative fits into the Web picture. IT managers must also understand how projects will be implemented and business objectives matched. Finally, C-level executives and business owners must understand how the Web site and online marketing strategies fit into the business as a whole, and quantify real business revenue and ROI.

Most books and Web firms cover an Analysis Phase based strictly on data (Web traffic reports, metrics, conversion rates, etc.). In version 2.0, XploreNet covers those important areas, but also delves in-depth, into the business reasons for a Web analysis, providing a case study at the beginning of Modules 1 and 2 to show the Web plan in action.

Understanding your Web strategies and implementing the right solutions is more than just data. It is about matching every part of your Web site, project tools, and Web strategies to your business and marketing objectives.

CASE STUDY

Our Water Utility Client, is a member-supported, international, non-profit organization that sponsors research that enables water utilities, public health agencies, and other professionals to provide safe and affordable drinking water to consumers.

Situation: Water Utility Client had a productive Web presence, but realized it was time to modernize the Web presence. After looking at 12 different firms, they chose XploreNet to help them facilitate collaboration and growth, positioning the Foundation as the world leader in research on water.

Impact: XploreNet put a Web plan together that included an intense Analysis Phase, reviewing Water Utility Client's past experience, defining the Web project's purpose and goals, understanding who should be involved, helping to develop the staff, defining success, and establishing ROI. XploreNet completely restructured the Water Utility Client Web site, making large portions of it database–driven, pulling content, not only from a database on the Web site, but also directly from Water Utility Client's proprietary database. The end result is a site, which is not only much easier to navigate but also delivers up-to-the–second, current content. The new site's subscriber feedback has been overwhelmingly positive. Now it is four years later and the site has grown to include knowledge management and online marketing tools.

Water Utility Client's ability to complete the Analysis Phase proved critical in the successful redesign of their Web presence and the ongoing improvements that have been added one module at a time.

1.1 PAST EXPERIENCE

Purpose: **Determine your current situation by understanding what is positive and negative about your current web presence.**

Key Points: **Putting your past experiences to work for you and starting fresh enables your new project to hit the necessary goals head on. Reviewing what went right and wrong during your last redesign will help the new project. You now have a more sophisticated understanding of the web and how your firm fits into the game.**

What was it like the last time you redesigned a Web project? According to *Kelly Goto,* the average time per redesign iteration would be a full-scale redesign and rebranding/repositioning effort every 24 - 36 months. After two to three years the finished product is typically not the one the client desired. Did it remind you of the last time you went in for a root canal? You've probably been through this process several times by now.

Did the following issues continue to occur?

➲ Long exhaustive meetings that produced few or no results
➲ Unclear objectives

- ➲ Unsure who should be in the meetings
- ➲ No task list
- ➲ Unsure who is running the show
- ➲ Unclear agenda
- ➲ Vague timelines
- ➲ No business objectives
- ➲ Confusion on technology choices
- ➲ Content soon defined as a "four letter word"

Water Utility Client had not been through such an intense process before. Their most recent Web site was completed by an internal Web developer who had put together a well-constructed project plan, but this was to be the organization's first step into building a complete and highly complex Web plan. Unlike many organizations, Water Utility Client's recent experience had been positive, so they went into the Analysis Phase with optimism and enthusiasm. Many organizations view this process as a nightmare. They often begin the new project nervous and uncomfortable, unsure of how the deliverables will be completed and if they will be completed on time and within budget.

LACK OF A PLAN

If you were fortunate and your redesign went through smoothly, count yourself lucky. Water Utility Client was one of the lucky ones, having an intelligent, well-organized, internal Web developer and a qualified Web vendor (XploreNet) to walk them through the entire Web project. For many firms, the reality is that most redesigned Web sites do not achieve a high level of satisfaction. According to Robert McGlouton, Vice President of SatisfactionWorks, a company that performs satisfaction surveys after Web redesigns, "Communication is an important key while understanding all the expectations of all parties involved."

The biggest obstacle is the lack of a plan. If the site you have isn't the one you wanted, then you did not have a proper Web plan. As the old saying goes "if you don't know where you're going, anywhere will do." Sure, you may have had a basic project plan last time, but you may have overlooked how the new site or application would fit in with all your other Web strategies. Water Utility Client now had to look at the business reasons, creating user profiles, and understand how their brand awareness would be affected by their new site. They were forced to match the business objectives, which included increasing the number of subscribers, improving value to subscribers, and improving online collaboration between consultants, to the Web strategies. Similar to getting that painful root canal, the situation can be seen more favorably if you know what the procedure is, you trust the team doing it, and you understand the anticipated outcome.

If you haven't used a Web plan before, now is your chance to get started. Put those bad experiences behind you and smile. Once you understand what went right and wrong the last time, you can make sure the next time is a successful experience, and even perhaps, a pleasant one.

1.2 YOUR PURPOSE

Purpose: **Evaluate your reasons for a new Web project.**

Key Points: **Understanding the purpose of the project and answering the simple question, "Why are we doing this?" will help a great deal. This is the section where you begin to understand the business reasons for the project and what will harm or help the process.**

If you're thinking about a Web site redesign or building a new Web application and you're having some hesitation, it may be because you haven't answered the first critical question: Why are we doing this? According to Net Magazine, firms redesign/rebuild their site every 15 months typically because the web is changing so quickly. Also, Net Magazine reports other factors like "accessibility, for one, and, of course, the time-honored tradition of corporate re-branding, which is as good a reason for redesigning a website as any other."

You can build some amazing backend (database driven) online tools due to the present advancements in technology. The problem is most development firms either do not know how to properly implement these tools or they're not sure which tool to pick out of all the options. If you understand your purpose, then a redesign can go well, very well.

Figure 1.2: Strategy Hierarchy

Water Utility Client knew their purpose, making the initial meetings and discussions straightforward. You may have many reasons, including leading-edge brand image, selling more products, qualifying prospects, improving customer service, creating more interaction, or improving internal communication. The purpose for your new Web project may involve all of these reasons, but have you identified them, put them in writing, and prioritized their importance? The purpose of Water Utility Client's new site was to make them the number one, best known, world leader in water research. Their purpose was written down, emphasized, and digested over and over.

THE PURPOSE

You've probably had a Web site for a while and now you're asking "What is the purpose?" You might have put up the first one or redesigned it a few times because everyone else did – competitors, vendors, customers, etc. When we've asked new clients what the purpose of the current site was they often reply "Everyone has a Web site." Did you ever ask why? Not only the purpose of the site, but why are you redesigning it? It's important for you to know. Keep reading and you'll find out why it's important for you to know why.

FIND OPPORTUNITIES

You need to comprehend opportunities that are either obvious or hidden. Water Utility Client identified three key competitors and during the Analysis Phase, learned to understand what made them similar and different. They strived to find out why a water utility would choose them over the key competitors. Follow Water Utility Client's lead and look at what some of your competitors are doing on the Web. Perform a basic competitive analysis. Is what they're doing making sense? Do they have some good ideas? Are they developing innovative digital ways to sell their products or services? In addition, look for the hidden opportunities by asking your current customers for their opinions (we'll get into this in more detail later in the book). As you identify opportunities and define their purposes, ask some key questions like:

➲ Will we increase the number of new clients?
➲ Will existing clients see an improvement in customer service?
➲ Is the redesign part of a new branding campaign?
➲ Is this the time to create more differentiation between us and our competitors?
➲ Are we launching a new product or service?
➲ Do we have a realistic budget?
➲ Will it help our employees to be more efficient?
➲ Are there alternatives to building the new application(s)?

According to Jacob Nielson of the Nielson Norm Group, "redesigned sites and Web applications have a low success rate and the many areas measured for improvement see only slight increases." There are many causes for a redesign to take place, but typically companies implement the changes just for the sake of change. They may briefly think about all the reasons listed in the first paragraph in this Module, but they don't have a real plan or understand strategies they are trying to accomplish. Change, for the sake of change, is not the best reason.

Water Utility Client's redesign was successful because their purpose was clearly defined and matched to overall company goals. In fact, the site won two APEX awards for best redesign and most improved internal Web pages. If you can *objectively* answer the "why" question then your next Web project has an outstanding chance of success.

1.3 REAL GOALS

Purpose: **Understand all the goals of the project and how they match your business and sales objectives.**

Key Points: **Evaluate the goals that matter – increased sales, revenue, profit, and cost reduction. After establishing the business goals, turn your attention to project goals. Write the goals down and then match them to the new project.**

How many firms put their Web goals in writing and perform a reality test on each one? According to the Franklin-Covey Organization, goals are over seven times more likely to be accomplished if written down on paper. We often hear prospects say "We don't need to know the details and/or flush out the goals; we just need a great looking site that increases revenue," or "just make it work". We frequently respond by asking a few questions of our own like "What if the goals are met with a poorly designed site that looks terrible?" or "What if the site has no unique tools, but quadruples revenue?" You probably look as confused as they often did.

Goals for a Web project or online marketing strategies do not have to be cumbersome or complex. In fact, often 3 to 5 is all you need. However, everyone on your team and the vendor (if an outside vendor is involved) need to understand what the goals are and more importantly agree that those are the key goals.

The following brief example will help clarify the point. The site www. google.com, does not meet any of the definitions of a flashy, interactive, or complex Web site. However, for Google the purpose is to get people into the search process as quickly and simply as possible. Their continued growth in both revenue and stock value suggests that their site and business objectives are working. The bottom line is that you must first detail the business and project goals in the Web plan, brainstorming on as many goals as possible, and then prioritizing them in order of importance.

THE SMART TEST

You may understand some of the reasons listed in the first section while moving forward with a redesign, but have you put the SMART test to them? The SMART test comes from the self-help industry and is used to measure goals. Are your reasons for the redesign or new application **S**pecific, **M**easurable, **A**greed upon, **R**ealistic, and **T**ime based? You should also add **C**lear, **C**oncise and **F**undamental (we'll call this CCF for now). If you can't plan then measure it, don't do it.

SWOT ANALYSIS

Water Utility Client came to XploreNet with eight site and project goals. They had spent time reviewing and understanding the importance of each goal. XploreNet helped them prioritize and flush out the details by performing a Web SWOT analysis (SWOT stands for **S**trengths, **W**eaknesses, **O**pportunities, **T**hreats). The strengths of your current site might range from a well-designed "look and feel" to quality content. Weaknesses typically range from bad design to a complete lack of calls to action. Opportunities like repeat sales can be uncovered when you complete the Analysis Phase of the Web plan. Threats are all over the place – what do your competitors' Web sites look like and how good is their content? For example, do they have a shopping cart and your site states "call to order"? The SWOT analysis will help you prioritize the key goals and eliminate unnecessary or ineffective objectives.

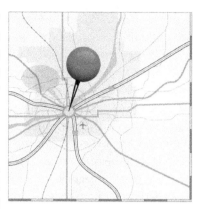

BUSINESS GOALS

Business goals include sales increases, cost reductions, brand evaluations and improvements, etc. Real business goals for a new Web project must be objective, quantifiable, and realistic. If you think that the new site will mean a 40 percent increase in traffic and revenue, are you thinking logically? How will you get there? What other strategies will be affected? How will you track and measure the results? Indeed, many firms have their heads in the clouds when it comes to the Web. Your goals may or may not be realistic so ask yourself as you're going through the process, "Are we in Fantasyland or is this truly possible?"

Some common categories for business goals include:

Increasing online sales
- ➲ The challenging part of this goal is that many firms do not have good records of their online sales. If you do, then what type of growth do you want to see? How many new visitors will it take to get there? How will the online store be built to achieve your goals? If the site visitor can buy online from you, then the redesign and accompanying tools must enhance that experience. Once you put the right tracking processes in place, the goals are easily quantifiable – number of new sales, increase in revenue, reduction in internal staffing, etc.

Increasing requests for information and/or appointments
- ➲ This goal could have a direct affect on revenue. The key is to understand the average lifetime value of a customer. These numbers should be quantifiable. For example, currently we're getting 10 leads per week and the new site will increase the requests to 20. By hitting our goals we'll increase revenue by 100 percent.

Increasing brand value
- ➲ If the focus of your firm is delivering offline services or products, then your site must enhance your marketing and public relations plans. This is difficult to measure; however, performing usability testing during

the development process to understand your brand value will create a measurement baseline. Then, once the site is launched, perform a follow-up usability test where the testing group again rates your brand. You may be able to identify something like, "We've improved our visitor satisfaction by 75 percent %," or "Prospects are demonstrating more brand recognition."

Improving customer satisfaction and/or decreasing internal staffing costs

➲ The belief has always been that a Web presence can improve the level of customer support. It can, if it's done underlined correctly. However, if your site must provide a high level of support to the visitor, then the tools on the site must be highly sophisticated. How do you measure the success rate here? Usability testing, online surveys, traffic increases to certain sections, or specific customer service tools are solid goal areas.

Publishing Information

➲ If your site publishes valuable information – downloadable articles, newspaper articles, training information, etc, you'll want to establish goals like increases in fee subscriptions, increases in total subscriber base, and/or increases in pay-per-download revenue.

PROJECT GOALS

Project goals go hand in hand with business goals and include budget restraints, timelines, team members, development issues, technology options, etc. Our Water Utility Client was different from most firms who typically have a single project goal. Our Water Utility Client had several project goals and used them to flush out the details of the scope of work. Project goals tend to be simpler to define than the business goals of the new Web project.

Example project goals include:
➲ Completing the project in three months, in six months, etc.
➲ Completing the project for a $100,000 budget, $90,000 budget, etc.
➲ Accomplishing weekly communication and updates
➲ New design complete in four weeks, in three weeks, etc.
➲ Database completed by March 1, or May 1, etc.
➲ 50 percent increase in traffic from Google, etc.
➲ 30 percent improvement in email click-through rates

PRIORTIZE GOALS

XploreNet helped our Water Utility Client determine the goals that mattered most to the organization. We helped them focus on three to four main goals and a few subcategories of goals. You may need to build the site or tool around the delivery of these points to the visitor. Clearly defined goals will assist you in understanding your main priorities and maintaining perspective as you run into project obstacles or challenges. Of course, you can't have everything. So, what will be most critical and what fits the long term company plans?

As most business owners and executives know, goals are always the key to success. No matter what the end Web goals, they must first be realistic, agreed upon, scheduled, and written out in your Web plan.

1.4 PERSONNEL INVOLVED

Purpose: **Understand what members of your staff need to participate in the project and what their project functions might be.**

Key Points: **This section covers the specific team members involved in your redesign or implementation committee. Understand which personnel should be involved, their level of involvement, and their project purpose.**

A true Web plan is put together by several members of your organization. Ever been to a party where no one talks? Or showed up for a meeting with all the wrong people? Web projects sometimes attract strange players to the table – some invitees and some not. Each participant on the Web team must have specific strategies and goals in mind. More importantly they must know why they are involved and what the bottom line expectation is of them. It has been XploreNet's experience that committees of 12 people can work out well, while two-person development committees may result in a fiasco. Usually when we begin a Web project, the client is often not sure who should be involved in the process. Assembling the right team and understanding how your team will work with an outsourced vendor <u>before</u> you get started can greatly affect the success of your next Web project.

TALENT INVENTORY

If you have an internal Web team, then you probably have most of your players available. You should have the following skill sets: strategist, marketing coordinator/communications director, project manager, front line employee, front-end development (look-and-feel person, graphics, etc.), and a technical person (HTML, complex programming, server issues, etc.). If you are missing team personnel or have personnel where they should not be, you're in for a rude awakening. You will likely end up with poor quality commentary, unfocused meetings, missing resources and above all, <u>lost time and money</u>. These same problems arise when you have too many people involved. We've seen committees of 12 to 15 people where tangents and alternative/hidden agendas take center stage. This is not an effective and efficient team.

If you're using an outsourced vendor, then access the talents of both your internal team and the vendor's team. XploreNet and our Water Utility Client created a project team as if it were a combined company. Both organizations understood each person's role in the team, clearly defined that role, and disclosed all expectations. We worked together to identify what talents were missing and how to fill the gap accordingly.

PERSONNEL PLAN

Our Water Utility Client was fortunate because they had an outstanding internal Web master and several willing committee participants. Their Web committee consisted of 10 people from different divisions with very specific goals in mind. Like Water Utility Client, if you have the right players (it can be two to 10 – try not to have more than 12 if possible) involved, the process runs smoothly and makes sense. The rule of thumb is that if you are a medium to large firm then you should have three to 10 people involved. They must be decision makers, who understand your overall company objectives, the services/products, and the core values. Typically this includes the President/owner, CIO, CTO, CFO, VP or Director of Marketing, VP/ Director of Communications, Sales Manager/rep, Marketing Executive/ outsourced Marketing vendor, office manager/customer service/operations, IT employee/outsourced vendor, and customer. You might have noticed customer at the end of the last sentence. Don't pass out, but yes it makes sense to have one of your customers involved in your web project. It's amazingwhat they will tell you.

If your firm is a small business, make sure you have at least two people involved. If you're solo, have a partner company, networking contact, vendor and/or customer involved in the process. If you can only have one other person involved, make it a customer. As was mentioned above, customers will tell you things you would have never thought of and often their feedback can be the difference between the success or failure of your project. In addition, a client will help you understand why they chose you and how your new Web presence can help attract more great clients like them.

The process of understanding who should be involved in the project and knowing which talents to use and when to use them can be complicated and difficult. However, if you take the time to understand the process and identify the key players, you then already have a 50 percent greater chance of success. The strategy meetings with Water Utility Client were productive because the right personnel were engaged at the right time.

PROJECT LEAD

Our experience has taught us that the project lead can make or break the project. Especially now that most web related projects—with the new sophisticated systems—require a level of knowledge far above intermediate. We've found that your project lead must have skills in three key areas.

First, the lead person must be technical by nature understanding both code and how servers operate. Second, this person must be "people friendly" - able to explain both opportunities and challenges as they arise. Third, they have to understand "look and feel" issues, complimenting all the back end tools with the right design elements. It is difficult to find someone with strengths in all three areas, but the success of your next project might depend on it.

1.5 DEFINE SUCCESS

Purpose: **Establish a definition of success from your specific viewpoint.**

Key Points: **In this section you'll want to write down what success looks like for your Web project. Use a broad stroke to paint the correct picture. Think from the end and be as detailed as possible.**

If someone walked into your company and told you that they were considering your product/service, but they also told you what information they thought was most important for them to decide to buy from you right now, instead of your competition, wouldn't you make sure that they received the information they needed? Most companies work hard to keep existing customers and to gain new ones. Yet how much time and money is wasted on addressing the *perceived* needs instead of the *known* needs of both existing and potential customers? Success to you or your firm may not be success to another firm. Before you get started make sure you know, in detail, what success will look like when your project is completed.

GOLD MINE

Almost every company with a Web site is sitting on a gold mine of information that could tell what their customers' true needs are, what is important to the customers, and what sales/marketing language is most effective for getting the customers to buy from them, instead of their competition. *Our* Water Utility Client had customer surveys and traffic reports filled with important statistics. They needed to put the information into a Web analytics format. Web analytics is a term used to describe the ability to measure and interpret information about how online visitors behave.

Here's an eye-opening statistic: recent studies have shown that on average, nearly 85 to 90 percent of potential purchasers of goods and services believe that going to a company's Web site is an important part of the decision-making process. If you have a Web site, what is it telling that 80 percent of your potential purchasers? If you don't have a Web site, or the site is poorly planned and constructed, how will that negatively impact the impression of your company? Will you be perceived as "real"? Think about all those potential customers looking at your Web site as they try to decide if they'll buy from you. Do you know what information is critical for them to make that purchase decision?

Let's think about your Web site's visitors in a more traditional sales funnel manner (see Figure 1.5).

Figure 1.5: Web Visitor Funnel

Web Visitors

50,000 total visitors to the site	
100% of visitors	Total Visitors
75% of visitors	Returning Visitors
50% of visitors	Product Descriptions
30% of visitors	Put items in cart
7% of visitors	Checkout process
5%	Complete transaction

196 visitors complete the scenario's objective

Most salespeople look at the universe and decide what type of individuals or companies might buy from them (suspects). Then, they ask themselves, who is more likely to need what I offer? By breaking down that universe into more specific groups or targets, the salesperson has just defined groups that are more likely to buy from them (prospects). From these prospects they focus again on those that have a more pressing need, and these people become qualified <u>leads</u>.

In the Web-enabled world, your Web site actually knocks out one layer of the traditional sales funnel. Everyone that comes to your Web site does so *voluntarily*. Rather exciting to think these visitors are specifically going to you to fulfill their needs. Somehow they have been enticed to sit down, turn on a computer, find your Web address, and look at your Web site. Why? Because they're likely considering buying from you and they want more information.

Now, let's look at the other side. Most visitors to your Web site know what they are looking for—on some level. Therefore your Web site should be clearly organized, copy should be concise and to the point, with standard terms used for the navigation (References, About Us, Contact, Pricing, Solutions, Services, Examples, etc.). Navigation should be obvious and consistent throughout the site.

So let's stop for a moment and check where we are. About 80 percent of the people thinking about buying from you will probably look at your Web site. The mere act of going to your Web site indicates that they have some level of interest in what you have to offer. The next question is, "Do you know what information they want?" This is the critical information that will compel them to buy from you instead of your competition.

MARKETING TOOL

Your Web site is an incredible marketing tool. Not only can you quickly change your Web site, but also those that visit your Web site are <u>qualified</u> leads. Here's a piece of information most people don't have: everything that happens on your Web site can be tracked, measured, and analyzed – this is Web analytics. *Our* Water Utility Client had been tracking some of the site users, but needed usable information. They had to determine how people found the foundation, what marketing/sales language was most effective, and what type of information their different subscriber groups (customer segments) needed to have to make a purchase decision. This is extremely valuable information; too bad most companies don't take advantage of such a gold mine. In Section 1.7 we take a look at how to access the information, how to interpret the data, and how to put both the information and insight into action. In the end, you'll have a definition of success and an established return on investment (ROI).

1.6 ESTABLISH ROI

Purpose: **Understand and articulate the ROI and tracking processes.**

Key Points: **Establish Web metrics that match sales and business objectives. Evaluate various Web tools to make sure your Web analytics are reasonable and simple. Create systems to quantify everything you do. Include as much data as possible in your Web plan.**

By now you've heard all the buzzwords around Web analytics, but what does it really mean? Although recent analytical software releases are improving the tracking process, most firms either do not understand how to use the tools or the additional features generate overkill. It is absolutely necessary to establish a baseline so you can track and measure online behavior over time. In fact, you must be able to quantify everything you are doing in order to get a true return on investment. XploreNet helped *Our* Water Utility Client establish a baseline to quantify success, the steps towards which success could be achieved over time. During the Analysis Process, *Our* Water Utility Client recognized how much a visitor meant to the organization in terms of life-time value and what the average cost would be to get the visitor to use the site tools. If you know that for every 100 visitors you get one sale at an average sale price of $200, you obviously want to improve the ratio of visitors to sales (sometimes called a conversion rate). So how do you do it? There are two categories for ROI – business objectives and user behavior metrics. Business objectives include sales figures, revenue goals, and profit targets, while user behavior incorporates Web traffic and statistical analysis. Another area of measurement is customer and partner feedback – their comments can lead to changes that directly affect the bottom line.

ROI

Return on Investment or ROI is a popular term these days. Have you sat through a vendor presentation where ROI wasn't discussed? Probably not, but did the vendor understand what would be measured to come up with an ROI? ROI is the central force for the success of any company, no matter how large or small. You will need to establish certain ROI goals for your new Web project. Establish breakeven points and points of no return (When do we stop investing in the site and look at a new strategy?). During the first strategy session with Water Utility Client, XploreNet discussed what would be the target for increased subscriptions and renewals (sales revenue). It was necessary to understand how much profit could be tied to the new project. We detailed these goals in the Web plan.

In order to study behavior, you probably already have the fundamental building blocks to understanding how and why your visitors behave the way they do. Every Web site visitor generates a history of the visit that tracks each page seen, how long each page was displayed, etc. All of these visits are compiled and stored in a Web site's log files. In essence, the log files are a historical record of how each visitor interacted with your Web site. Your Web host probably has a few months of your Web site's historical log files. These are usually saved on a month-by-month basis. Many hosting companies offer monthly reports that show standardized reports regarding your Web site. This information varies based upon the Web analytics service your Web hosting company subscribes to. But, almost always, you'll at least have the following information:

- ➲ Total Visits
- ➲ Average Visits per Day
- ➲ Average Page View per Visit
- ➲ New Visitors
- ➲ Total Page Views
- ➲ Average Page Views per Day
- ➲ Average Hits per Visit
- ➲ Repeat Visitors

- ➲ Total Hits
- ➲ Average Bytes per Day
- ➲ Average Length of Visit

USING THE DATA

Now that you have the information, how do you put it to work? One month's information doesn't mean too much. What really counts is the trends established and verified in the month-to-month reports. *Our* Water Utility Client's three-month data trends showed that most visitors used the pages found in the project center. If you notice that your total number of visitors is declining, and the average length of the visit is also declining, check your sales. Often, sales will have dropped within the same time frame.

This leads us to interpreting and applying the information from that data. If you're fortunate enough to have access to more comprehensive reporting solutions, here are some of the areas that you should consider investigating:

- ➲ Top path analysis
- ➲ Key search words/phrases
- ➲ Referring URLs

Web analytics or metrics are industry "buzz words" to describe the process of quantifying Web activities and objectives. Some real world examples may help you see how you can use this information:

Example 1: You use your Web site to showcase your work and generate sales leads (your Web site has two goals - show examples of your work and have people contact you because they are considering using your services or purchasing your products). Using Total Page Views, Average Page Views per Visitor, New Visitors and Repeat Visitors, let's look at how these numbers trend over a 3-month period.

Do the numbers for each stay the same or fluctuate? More than likely, the numbers change month to month:

- ➲ Total Page Views: Total Page Views will give you a rough idea of how much activity is on your Web site.
- ➲ The Average Page Views: The Average Page Views per Visitor indicate how involved your visitors are with your Web site. If the number is less than 2, it could mean that most visitors are confused/bored/uncertain after arriving at your homepage and don't know what to do next (this assumes, of course, you have at least 10 pages to your Web site). Most visitors see, or don't see, something that makes them leave your Web site.
- ➲ New and Repeat Visitors: Let's add New and Repeat Visitors. Do you have more New Visitors acting this way? If so, something on your Web site within those first two pages is driving them away from you, make

sure your links work, your copy is clear, and you have a defined path/ goal for visitors to take next (e.g. click here for examples, click here for more information, click here for references). More than likely, your visitors don't know where to go next. It's your job to lead them.

Example 2: You sell products or services online. By analyzing information from the Navigational Path report, key search words/phrases, number of visitors, and your total sales, you can figure out who (Google, Yahoo, MSN, etc.) is bringing you the best traffic. Here, we define best traffic as those who buy from you, not just those that visit your Web site. By working backwards, we look at the total number of people that saw your "Thanks for your purchase" page (from the Navigational Path report). We then see the pages that most of those people viewed before they saw that page. All we're doing is recreating the most popular pages viewed (and in which sequence) on your Web site. Notice which pages people viewed to arrive at the "Thanks for your purchase" page. Here are the Web pages that directly impact your sales. What is on those pages? That information is crucial in driving your visitors to buy from you.

Figure 1.6

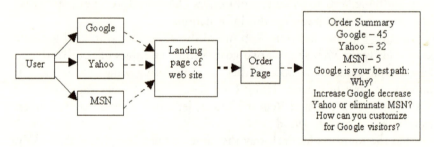

Example 3: Using the same report, look at the top paths. How long are they, and where do they end (at what page)? You may be losing visitors along the way to your "Thanks for your purchase" page because of something you're not providing. Look at the pages and see what topics they discuss, what those pages have in common. Is the copy clear? Are links working to the next page? Have you clearly outlined the next step you want your visitors to take? We keep looking back through the sales funnel until we arrive at your home page.

Establishing ROI is critical to any Web project. You will need to understand some basic metrics set up and then match your business goals to the metrics. At this point, you will realize your Return on Investment (ROI).

1.7 CHOOSING A FIRM

Purpose: **Develop a game plan for choosing the right Web firm.**

Key Points: **Understanding the best fit for your firm's size, goals, and budget is of critical importance. What are the characteristics that make sense? What type of firm is <u>not</u> a good choice?**

Now that you understand the overall objectives and you have the right players in place, it's time to choose the development company. Choosing the right Web firm is critical to your company's Web project success. Choose wisely. The company you choose can either make or break your Web plan.

There are probably 10,000 or more Web firms in the United States, divided between large enterprise integrators, small to medium sized Web shops, advertising agencies, and numerous independent contractors. Picking the right firm from all these choices can be like playing Russian roulette. *Our Water Utility Client* simplified the process by preparing interview questions with a clear understanding how the chosen firm would help the organization complete its custom Web plan.

FIRST TIME

If you've never worked with a Web firm before, you need a Web plan. If you've never worked with a Web shop, the cost estimates you receive may give you sticker shock. In fact, you may be tempted to find a low-cost, independent developer. If you've had a good experience with this type of vendor, then count yourself fortunate and keep using that vendor. However, if you've gone this route, you were more likely to have had a bad experience. You've probably seen the bad and the ugly. Picking the right company is about matching your strategies and project goals to the firm that has the best chance of delivering.

DO YOUR HOMEWORK

As you go through the process of picking the right Web firm, **do your homework**. We recommend you understand the following before you engage a firm: (Most of these questions are answered in Modules 1 and 2 of your Web plan.).

- ➲ What do you want to accomplish with your Web site?
- ➲ What are your goals for this project?
- ➲ Why will your visitors benefit by coming to your site?

- ➲ How often will this site be changing? (i.e. content, structure, graphics)
- ➲ What makes your company competitive?
- ➲ What does your company do?
- ➲ What will determine if the Web project is a successful investment or not?
- ➲ What are your goals for this project?
- ➲ What is the level of computer experience of your audience?
- ➲ In your opinion, what makes a good site and what makes a poor site?
- ➲ Do you have a budget set aside? How much have you allotted for each strategy?
- ➲ What is your Internet experience?
- ➲ Do you have a budget set aside for monthly hosting and maintenance?
- ➲ Do you have graphics that will need to be developed?
- ➲ How much of your content is completed?

RECEIVING REFERRALS

As we mentioned before, there are numerous choices in Web firms. So where do you begin? Your best bet is to begin with a referral. Ask associates, friends, professional contacts, etc., which firms have done a great job for them. *Our* Water Utility Client found XploreNet by asking the company that performed their offline marketing for a referral. When you have identified a few (typically three or four is sufficient) take them through the full interview process.

INTERVIEW PROCESS

Water Utility Client looked at 12 companies and interviewed four of them by organizing two different question and answer sessions. They wanted to make sure the prospective firms and their team could ask questions of one another, thoroughly evaluating which firm would be the best match. XploreNet asked project and business-related questions that helped our team assess where they were in the process and how we could take them from point A to Z (most questions will be easy for you to answer if you have done your homework).

Once we understood their business, they needed to know what we could do for them. *Our* Water Utility Client understood the technology options, XploreNet's approach, XploreNet's methods to get more value for their budget limitations, and the best type of implementation. According to Istudio, you should ask to see examples of a consultant's successful Web projects and attempt to answer the following questions:

➲ Does their own Web site match their objectives?

➲ Do their case studies match some of your needs?

➲ Are there errors on their site or broken links?

➲ Is the content on their site relevant?

➲ How many projects, similar to yours, have they completed?

➲ Do they have information about their processes and does it make sense?

➲ Who are their team members?

➲ Do they have the experience to help you?

➲ How did you find the firm?

THE BEST MATCH

After taking the prospective Web firms through the interview process, look for the following information from their answers. Look for the firm that:

You can create a relationship with
▶ Ask if they have relationships with their clients that are more than three years old.

Has experience with projects having similar goals to your project
▶ If your goal is to create a site to develop additional sales channels, ask what their experience is in this industry.

Ability to handle complexity with multiple software platforms, vendors, and potential problems
▶ Has the firm been to war, battling all the potential issues that come up when multiple products/software packages, and systems are involved. Can they demonstrate how they handled the problems?

Has a strong outsourcing/referral model
▶ Don't buy into the one-stop shop. It doesn't exist. The company that "does it all" does none of it well.

Has been in the Internet industry for at least five years
▶ How much can you afford to pay for their inexperience?

Has a clearly defined process for Web projects
▶ What are the different steps, from beginning to end, to complete a project? What makes it successful?

Has a real business location
▶ What happens if your developer is not committed to the business? Having a business location shows a strong level of commitment.

HELP WITH YOUR WEB PLAN

Your most important step in choosing a firm: determine how they will assist you in developing your Web plan. XploreNet came to *Our* Water Utility Client with a Web plan outline and showed how the project would be successful upon implementing the strategies of the plan. If you don't have a Web plan, the firm you work with should help you implement one. If you have a detailed Web plan, match your strategies to the best firm.

1.8 STRATEGY MEETINGS

Purpose: **Understand how to organize and run a successful strategy meeting.**

Key Points: **Understand the agenda for the meetings and what the end results should be, upon completion. You should have a plan going in to the meeting and know what areas will be potential stumbling blocks.**

According to Mike Ewart-Smith of the Whitegrove Group, "If the formulation of strategy is not being undertaken in strategy meetings, there is probably no strategy at all." A strategy meeting for your new Web project should cover the business strategies, business goals, project goals, and other key components discussed in previous sections of this book.

XploreNet guided Water Utility Client through four strategy sessions totaling 14 hours. We utilized our experience of conducting more than 200 strategy sessions, helping Water Utility Client detail the agenda, and hit the most important priorities. Our unique agenda format helped them create an effective Analysis Phase plan. Your strategy sessions will be successful if you have a proper agenda and understand some of the "dos" and "don'ts" of the meeting format.

AGENDA

First, establish the agenda. Make sure you consider *all* the opportunities and issues that must be covered, in order to build the foundation for your project's success. This includes not only the main goals, but also all the different departmental issues. If you are a large firm, you will need to include several areas. If you're small, the issues may be smaller. Give the meeting a catchy name that will excite the participants, but make the purpose and goals of the meeting apparent. The tag line of the meeting should be something interesting, challenging, and significant. Accentuate the importance of the meeting and persuade participants to buy into the excitement. For example: ACME Company Web Strategy Meeting: Developing the Path to Success for 2005.

If you are to head the meeting with your own team, communicate with the lead consultant of the development company on what you want to see covered and the key strategies that should be defined by the end of the meeting. Make sure you gauge the time necessary to cover all topics – you may have to organize two or more meetings to completely cover every detail. Next, create a list of all participants and ask them what topics they would like to cover. Emphasize the importance of the meeting and why they must participate.

At the end of the book we provide a strategy session guideline. Be sure you prepare copies of the agenda in detail, being careful not to include too much information, but enough to accomplish the priorities. Create a statement of purpose and an overview of steps to illustrate to the participants how the team will get from A to Z. After this meeting, you will know what you want to cover, the necessary time commitment, what materials to use, and who will be involved. How do you make the process work?

DOS AND DON'TS

The key question becomes, "How do we run a successful strategy session?" Here are a few tips for the process:

- ➲ **Do** utilize flip charts and hand outs
- ➲ **Do** have some type of tool to illustrate the flow chart of the project
- ➲ **Do** make sure you have a high speed Internet connection
- ➲ **Do** use PowerPoint via a projector or web based system (like gotomeeting.com)
- ➲ **Do** plan for a short break, once every hour
- ➲ **Do** tell participants to turn off cell phones, PDAs, etc.
- ➲ **Don't** run over the scheduled time period
- ➲ **Do** introduce the team members (unless every one knows each other) and discuss their reason for attending the meeting
- ➲ **Don't** let other issues get onto the agenda -- create an "other issues funnel" (meaning this is where issues that fall outside of the agenda will be listed and can be taken care of outside of the meeting or at the next meeting)
- ➲ **Don't** let one particular agenda item take more time than first estimated. If the issue needs more time, delegate the issue to someone who can take care of it outside of the meeting
- ➲ **Don't** let a few participants monopolize all topics

➲ **Do** take notes and record an audio of the meeting
➲ **Do** summarize at the end of the meeting
➲ **Do** set up the parameters for the next meeting

A successful strategy session should cover all the topics we have covered in Phase 1 and begin laying the groundwork for Phase 2.

1.9 CREATING A CREATIVE OR TECHNOLOGY BRIEF

Purpose: **Understand the key elements of a creative and/or a technology brief.**

Key Points: **In this section you will learn the elements of a creative brief and a technology brief. You will understand how to write one and what it should and should not include.**

A creative brief (in our industry often referred to as a technology brief) is a written overview of your project from separate angles. Typically both the internal team and web vendor/ad agency are involved in the process. XploreNet assisted *Our* Water Utility Client in completing a creative brief, covering background, backend features, "look and feel" issues, user goals, usability, navigation, etc. The technology section helped Client with the technical options: technical skill levels, technical platforms, backend options, API choices, and hosting environments. This type of brief describes the goals of the project, just as the Web plan describes the Web strategies. When properly developed, the brief will enhance the scope of work and enable the client/team to fully understand the client's requirements and necessary execution.

A great brief will result in quality production, no misunderstandings, and client satisfaction while a poor brief causes heartache, wasted time, and unworkable options that lose creativity and sustainability.

CREATIVE BRIEF

Taking a closer look at the creative brief, the document outlines the objectives, perceptions, positioning, audience, communication strategy, and assumptions for the project. The creative brief details the creative concepts the development team intends to implement. What does it look like?

We have included one in the Web Plan in Module 5 as a template for your use. The first part of the brief covers the project summary, followed by general project information, goals, and relevant background information for the Web project. The initial paragraph should be a statement overview of the project as a whole. This part is similar to an executive summary and covers both short and long term goals.

The next section covers perception/tone/guidelines, and facilitating how your target audience should respond to your new online presence. This paragraph covers the details of what the target audience thinks and feels about your company and the site. In addition, you will need to detail what you want your visitors to think and feel. For example, it should include adjectives that describe the way the site and your company should be perceived by the target audience. Finally, you should cover the specific visual goals the site should convey.

The third part includes the communication strategy. At the end of this section you will understand how to convince the target audience to make a purchase decision with you. Important questions include the overall message you are trying to convey to your target audience. How will you convey the overall message? It is at this point that you will need to identify the stages of development (if appropriate) used to execute your goals.

Lastly, communicate how you intend to measure the success of your Web project. This section covers an evaluation of the existing site to compare to the new one and the competitive positioning. The final section deals with the targeted message. You must answer the questions, "How are we different from our competition and what factors will make us a success?" Detail the factors that make you different and what specifically sets you apart from your competition. You must state a concise word or phrase that will appropriately describe the site once it is launched. This should include the calls to action and overall brand the user experiences.

TECHNOLOGY BRIEF

The technology brief is often overlooked. This document details the technical options and processes the development team intends to utilize. A technology brief fills in the gaps that are left when you only use a creative brief. The additional information includes technology options and the process necessary to implement the creative brief. In other words, how will the technical or functional side of the project work in conjunction with the creative side? This is where you create a balance between cool, neat, new technology, and the user's ability to use it.

The technology brief defines the platforms to be used and the skills required to use them. It also defines the functional requirements. Many sites and applications look great but are useless to the users. Developers frequently overlook how the graphics and content will fit in properly with the functional requirements.

The creative brief and technology brief are key components to planning your project and they must fit properly in the blue print section of your Web plan.

Please e-mail byoung@xplorenet.com for an
example template or refer or log on to www.
webplicitybook.com
Use the code web20.

COMPLETION OF MODULE 1

The Water Utility Client project and resulting Web plan was successful because the Analysis Phase covered in Module 1 was well organized, detailed, and complete. In fact, SatisfactionWorks completed a survey of the Water Utility Client users after the redesign and found that the redesign achieved a 93 percent approval rating – the highest rating they had ever seen.

The Analysis Phase is different for every firm. If you've dedicated the time to complete the analysis, you are well on your way to a successful Web project. Be sure to refer to Module 5 for the outline and details of a properly constructed Web plan.

"I have seen the vendor portals that you have designed and I must say we are impressed. I hope the rest of this project goes as smooth and as comprehensive. Thanks for your hard work. "

Executive Director
Cynthia Banks

MODULE 2

BLUE PRINT

2.0 Blue Print

This module discusses the next phase of understanding how your Web projects fit into your Web plan. This is the point where the Web strategies begin to take life and meet in the middle, with the project goals. Once you have completed this module, you will understand the details of your project. As you learn about blue printing, you will quickly understand whether your Web strategies are on life support or whether they are dead. Companies that successfully implement the correct blue print typically build a truly great Web project. This is where the theories and ideas become real. The blue print should detail the anticipated number and style of visitors, personalization issues, relevant calls to action, the scope of work, technology options, content requirements, initial design thoughts, usability, technical personnel involved, and future technology plans.

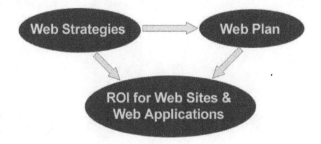

"I think you will agree that XploreNet did a great job in matching the design, page layouts, and functionality specifications that you had previously approved. None of the competition has a site that so adeptly merges form and function, identity and usability. Once again, I think we've all collaborated to raise the bar. "

Baraka Creative
Steve Reilly

Most modern Web sites do not meet the criteria of *effective* Web projects because they did not have a good blue print. They completed pages without understanding the purposes of the finished product.

Most books and Web firms cover a Blue Print Phase strictly from their point of view, completing a scope of work. This is an important step in the process; however, you must understand who is using the site, how they use it, and what you can do to keep them using it. We suggest that the blue print serve as your reality check. This process enables the executives to connect their strategies with the finished product and the marketing personnel to make sure their initiatives are being met.

CASE STUDY

Our Software Client is a leading provider of Six Sigma training, support and consulting services designed to help organizations achieve breakthrough level performance. Software Client prides itself on its top quality training programs for various training levels in the six sigma process and all levels of management. Training classes are available through public classes and on-site programs as well as through our client's web-based platform.

The existing web site did not meet the credentials of an industry leader. The company's VP of Marketing realized that the Web presence was a key component in the company's continued growth and success. The Software Client looked at several Web firms, narrowing the list down to four finalists and ultimately choosing XploreNet to implement their new Web plan. They chose XploreNet because of our combined experience in both Web site graphic design and database work. Furthermore, XploreNet's depth and breath of knowledge in building complex, database-driven Web sites showed. Software Client that XploreNet had the ability to help them develop a site that met their growing ecommerce and brand building needs, concurrently.

XploreNet put a Web plan together that included a detailed blue print, creation of a user profile, flushing out calls to action, completing a detailed scope of work, confirming technology options, organizing content areas, performing usability testing, and building towards future growth of the company.

. Software Client ability to complete the Blue Print Phase proved critical in the successful redesign of their Web presence. Their blue print has been used to incorporate online marketing tactics and back office tools into their overall business plans.

2.1 UNDERSTANDING VISITORS

Purpose: **Understand the visitors that will use your Web project once it's complete, whether they are internal or external to the operations of your firm.**

Key Points: **In this section, you will learn to understand how your visitors will utilize your new tools and make logical assumptions as to their behavior, whether they are external (customers or partners) or internal (employees).**

Imagine a firm that offers custom homes, but delivers the blue print on the first appointment. When it comes to the Web, many firms are getting bogged down by this problem. Our Software Client knew that their site was not appropriate for their target audience. But they needed to have a better understanding of how users utilized their site and Web tools. They felt the lack of feedback data on their users meant that the existing site did not meet or exceed the user's expectations.

The Software Client site was built mostly from the company's viewpoint, without a thorough blue print. The blue print only included the company's viewpoint, not that of the customers. This is a common problem for many firms. According to Forrester Research, only 33 percent of companies currently improve their online operations by making use of the data associated with how customers use the company's Web site. This means that nearly two-thirds of the businesses do not go into Web projects with the correct information.

USER BEHAVIOR

Professional Web-based marketing companies know that user behavior is more than just data. Simply measuring page hits, page views, click-through rates, and conversion rates is not sufficient. Our Software Client had good traffic reporting data, but it was critical for them to have a quality understanding of their user's preferences and purchasing behavior. As most consultants agree, most traditional businesses have proven that understanding customer behavior is a key priority in keeping a customer. In fact, it can cost five to ten times more to get a new customer then to keep a current one. So how do you avoid losing existing customers while still attracting new ones? The reality is that you need customer information to help understand what's happening with your Web presence. You need demographics, buying patterns, technical skills, and personal user preferences.

USER PROFILE

During the Analysis and Blue Print phases of the Software Client project, we created a user profile that described, in as much detail as possible, the general demographics and buying tendencies of their users. We strived to answer the question, "What do the Software Client customers want from the site?" For example, Software Client profile included middle managers to executive level, tending to be male in the 35 to 50 year old age range, high income, and very comfortable with the Internet. Additional demographic data might include marital status and Internet habits. In addition, you will need to have a good idea of how often and when users use the site, based on the review and establishment of metrics.

Now, what do they want from you? Later in this module we'll discuss usability, but at this point we're attempting to set up the parameters for the usability testing. You need to answer questions like:

➲ Do they want to make a real-time online transaction?
➲ Do they want to review and edit their customer data?
➲ Do they want to schedule an appointment?

Software Client wasn't sure what their users wanted from the site. This is a common problem because it is difficult to truly gauge what, exactly, your users want from your site. The findings can be subjective. However, you can make an attempt to design your ultimate user.

Typically users fall into three to six groups. Our software client identified three key groups. You may need to do some "digging down" to find multiple levels of users. Software Client user profile helped develop the plan for design and navigation changes. Indeed, one of their goals was to uncover the best way to motivate the site users to take the desired actions. For this, they needed answers to the following questions:

➲ Who is coming to the site?
➲ Who are our most important customers?

- ➲ What will they do once they get there?
- ➲ How do we receive them?
- ➲ What denotes their purchase behavior?
- ➲ Which online tools will generate the most profitable customers?
- ➲ Which products or services increase cross-sell and up-sell opportunities?
- ➲ What modifications will make our new Web project more customer-focused?

IDEAL CUSTOMER EXPERIENCE

Every user has a certain level of expectation when it comes to customer service. But how does a user define the ideal customer service experience on your site? When customers are utilizing the Web, it's on their own terms, but knowing what they want is crucial to your company's success. According to user research, there are five key actions in the user process:

- ➲ Collecting information to make the right purchasing decision
- ➲ Identifying their own needs
- ➲ Identifying the correct source to buy from
- ➲ Negotiating the transaction process
- ➲ Evaluating their own satisfaction level with the transaction

Our software client needed to understand what motivated the user and pushed their hot buttons. Why would they pick the Software Client over many other competitors? Don't overlook the fact that just because your site or application works correctly, orders will come in.

Understanding user's behavior is difficult, but not impossible. Set up the parameters and understand the tendencies, then do the best job possible to fit their needs. Software Client established a detailed, quality, user profile and the ideal user experience. This was an important step towards understanding the user's behavior on the site and creating a personalized experience.

2.2 PERSONALIZING YOUR WEB PRESENCE

Purpose: **Understand how to personalize your Web presence or Web application to each user's unique tastes.**

Key Points: **In this section, you will learn how to personalize your Web project for specific users. Based on how your users will use the site or tools, implement personal preferences.**

Personalized Web sites have come about as a result of the natural progression of the Web from a push process, to a pull process. The Internet has matured and users want information on their own schedules and preferences. The challenge for any Web site owner is to match the individual tastes of each user to the site's offerings.

Youtube.com is a great example of the user customizing their personal experience. They define how and when videos are uploading, providing ratings and feedback on the quality or lack thereof for uploaded content.

Can you really match each visitor's preferences when they visit your Web site? The reality is that you can match various user groups to specific offerings and in some cases personalize each user's experience.

PERSONALIZATION DEFINED

Personalization on the Web can be defined as providing a Web experience where the user's unique tastes and actions are customized. The experience can be something as simple as providing the right search tools to something as complex as providing all current and past account activity. The actions can range from simply providing the right "look and feel" to foreshadowing the user's next moves.

Web personalization includes customized Web pages that categorize users and matching pages to actions that make recommendations to the user. The ability to provide personalization on the Web comes from not only understanding users but also from identifying which sections or pages users visit on your site. It is now possible for you to personalize your product message for individual visitors on a large scale. This is accomplished by tracking user behavior on your site, down to specific, single mouse clicks.

IMPLEMENTING PERSONALIZATION

The methods for accomplishing a personalized Web project are user-defined preferences or group assumptions. Today, there are numerous new technology choices to help access and implement the right personalization tools.

The typical process is where the user-defined preferences take place immediately when the user first hits the page and signs in. The user inputs information and preferences and then receives a username and password (For example, Amazon.com performs this process). When the user returns to the site, information that has been previously stored is then delivered. Only the pages and information important to the user or that match the user's previously entered guidelines are then displayed. This type of approach requires the user to give up some privacy and costs more to develop.

Group assumptions are based on the known data of typical site users, like traffic reports or product preferences, so the site owner can provide information to those groups. Assumptions can be made about what your typical user will request and what the next steps might include. Furthermore, you can provide a path to specific information by placing yourself in the position of the user: "How would I buy from you?" "What would be the next common-sense click?" This approach requires in-depth research to understand each group of preferences but is less costly.

Personalized sites and applications help to establish an experience with a human element. By providing personalization, you prioritize the user first in your approach by listening to their wants and needs. By understanding how to fully utilize personalization in your Web project, your Web site will be judged as "user friendly".

2.3 CALLS TO ACTION

Purpose: **Understand how to persuade your users to take your targeted action.**

Key Points: **In this section, we continue creating the blue print by analyzing what a proper call to action is and how to implement it. The key is to have a balance between providing what the users want and ensuring your company's goals and objectives are met simultaneously.**

You are probably familiar with the term "call to action". If you've had any prior sales training or been in business for any significant amount of time, you know what it is and how important it is to the life of a product or business. A properly placed call to action on a web page can drive thousands of additional dollars to the bottom line. In fact, on the Web "calls to action" become even more critical to the success of the new Web project. An overly zealous process can frustrate users and drive them away while the lack of proper tools leaves the user baffled and moving on to your competitor's site.

The Software Client Web presence, like most firms on the Web, did not clearly define the calls to action. In fact, Forrester Research estimates that 45 percent of corporate sites lack a call to action or have a call to action that is confusing. However, they knew how important the right action was to their client and were willing to make the necessary adjustments.

DEFINITION

A call to action on the Web is the statement or action that directs the user to complete your desired action. Assuming that you have a standard home page, the call to actions must be prominently placed in the upper half of the page because Web users are notorious for using different screen resolutions and monitor sizes.

In other words, you must provide the appropriate information, at the right time, to enable the user to follow your guideline and ultimately make the ideal decision. The Software Client user groups ranged from individuals signing up for classes to executive-level decision makers looking for an enterprise software solution. The calls to action for these groups were very different.

You have probably visited a site and wondered, "What do I do next?" You may have interest in the product or service, but you're not sure how to learn more or complete the transaction. Can you imagine going to the Airport and not seeing any signs or personnel? It would be very difficult for you to get to the right plane at the right time. The software Client users had different needs, so the calls to action needed customization, directing the user to the proper process, based on the required action. In fact, most users are left confused on many sites and typically leave because the site lacks the appropriate calls to action.

IMPLEMENTING CALLS TO ACTION

How do you implement the right calls to action? First, match the actions to the goals. Review the goals you established in the Phase 1 Analysis step to come up with the appropriate ones. For example, your goals might be to increase revenue by 25 percent and increase the number of Web inquires by 15 percent. How do those goals match the calls to action?

Second, define the calls to action. The software Client realized three key calls to action: (1) enterprise software prospects were to call the company, (2) those visitors searching for training were to sign up directly for the e-learning class of their choice, (3) if the visitor failed to fit into one of the first two groups, that visitor would then be directed to buy products from the online shopping cart. The most obvious calls to action are: "Purchase Now," "Add to Shopping Cart," "Understand Details," "Download White Papers," "Subscribe," and others. More descriptive calls to action include: "New Visitors Start Here" or "Returning Customer". The software Client had to get the proper call to action in front of the user as soon as possible. If their visitor could not immediately locate where or how to purchase, sign up for the newsletter or find help, they could be instantly lost for good.

GUIDING THE USER

By now, you've defined what the calls to action will be. Now, it's time to explain your process to the user, answering the question "How do I do business with you?" Software Client had to describe the calls to action on each page and tie each to the proper content. They would increase the satisfaction level of the user by referencing their appropriate action steps frequently.

In your situation, the call to action may be as obvious as, "This is how we do business" or "New visitors start here." You're telling the visitor to "do this" or "take this step next". For example, on the main page, include text that tells about your services and then say "for best results, fill out our Contact Us form". Then provide a link to the form from that content. You create trust and provide real customer service when the calls to action are combined with assurances like "we never sell your information" or "we offer a 100 percent money back guarantee". In the end, you're providing as much help as possible to the user so that they can do business with you.

Calls to Action are critical to the success of any Web project, but particularly for a redesign. Our Software Client wanted to guide the user from the time the site was entered to the completion of one or all of the call to action steps.

2.4 DEFINE SCOPE OF WORK

Purpose:　　　　**Understand all details and sections of a properly developed scope of work.**

Key Points:　　　　**In this section, technical meets strategy. A clear and concise scope of work requires strong attention to detail. The scope of work document will be used to manage and implement your entire Web project. It is a combination of the information obtained in the Analysis phase and the answers derived from the creative brief and technology brief.**

A scope of work is similar to a blue print for building a house. Often, it is referred to as a statement of work, working document, or project guidelines. This document is used to determine whether your ideas and goals are being properly developed. It should cover the details of design, navigation, communication, and development. This document is your Constitution for the project or customer Bill of Rights. It states how, who, when, why, where, etc.

The Software Client came to XploreNet with a well thought out and detailed RFP, identifying the project overview and their thoughts in a rough draft scope of work. XploreNet's responsibility was to help the company develop, confirm and enhance the scope of work to fit into the overall Web plan.

PROCESS FOR DEVELOPING THE SCOPE

We're sure you've heard of the problems associated with using a bad contractor for a bathroom remodel or room addition. These problems can often be traced back to a poor plan or blue print. In these cases, neither party has an identical image in mind when it comes to the details and what the final product will look like. Whether your project is complex or simple, you must have a thorough, defined scope of work.

When is the best time to draw up the scope of work? Software Client was fortunate because they had a VP of Marketing who understood the importance of having some initial thoughts on the scope of work before the process began. The process for fully defining the Software Client scope of work began during the Analysis Phase and was completed during the Blue Print Phase. XploreNet conducted two meetings, even though one meeting is usually sufficient. During each meeting, the question and answer session helped lay out the scope of work and solidified the Software Client project blue print. The Software Client did not confirm the scope of work too early on. This mistake allows functional or technical requirements to be overlooked. In fact, if you wait until pages are being created before the scope is finished, then you'll have a misunderstanding regarding the finished product.

TYPICAL SCOPE CONTENTS

The scope of work should include deliverables on the technical side, but also include content categories and check lists. Bullet points and flow charts cover the functional aspects of the site, while content includes key phrases, paragraphs, pictures, and processes. A correctly crafted scope of work can help all personnel involved understand the details. The lack of a fully developed scope of work can cause great frustration. The scope should contain as many design and technical details as possible. A scope of work typically covers the following areas:

- Goals
- Deliverables
- Flow chart
- Assumptions
- Methods for handling problems
- Content check list

- Approach
- Timelines
- Personnel involved
- Project management process
- Challenge and Change Management
- Cost

The goals

➲ This is very straight forward. In the Phase 1 Analysis stage, you defined success and detailed the goals. Make sure you include both business and project goals. Write them down. They can be general, (four to five areas) or each and every goal can be defined in detail.

Deliverables

➲ The deliverables include all bullet points of the project and should be broken down into several areas: strategy/consultation, design/user interface, development, programming, and online marketing tactics and tools. This area should specify exactly how the database tools will be built and how they will work. In addition, describe all sections and the content required for each section.

Flowchart

➲ The flowchart should visually demonstrate the method by which the pages and sections will flow. The basic example of this is a site map. However, a flow chart also demonstrates all the steps and paths a user can, and/or will take. Match your calls to action to the navigational flow.

Assumptions

➲ These sections are straight forward. In this way, you will not assume too much. In a scope of work you want some assumptions. These include seemingly trivial pieces of information like the size of pages, the type of technology used, what is not included in the scope, and the financial commitments. Examples of some assumptions are that the site will be built for maximum efficiency for DSL connections or that the client is responsible for all merchant and licensing fees.

Content checklist

➲ This list includes all areas of content, what is required, what is missing, and/or what needs to be created. Content includes all text, pictures, graphics, and database schemes.

Approach

➲ How is the internal Web team or outsourced partner going to accomplish the deliverables? The approach should detail step by step how each part of the project will be completed.

Methods for handling potential problems

➲ Make sure the scope includes a method for overcoming challenges. For example, during many projects, new ideas come up that might be outside the scope – how will they be handled? Who is responsible for bringing up problems? A good scope of work details both the challenge and change management of the project. This section defines how the issues will be handled before they come up and who will be responsible for fixing them. It also helps with scope changes. Changes are a part of every scope of work and the method for discussing, pricing, and implementing the changes should be covered <u>before</u> you start the project.

Timelines

➲ A critical part of any scope of work is the timeline. Define the established milestones for the project and to whom the responsibility falls for each one. This should include *both* the client's and vendor's tasks and deadlines.

Personnel involved

➲ This section should include the names, contact information, and detailed responsibilities of all personnel involved in the project. If you are using an outsourced vendor, this section should also include the vendor's background and expertise.

Project management process

➲ Detail how the project will be managed in this section. Ask and answer the questions: "Is there a set process?" and "What tools will be used to monitor this process?" Also, will the client need a user name and password to the project management site?

Challenge and Change Management

➲ Every project has challenges and often you don't see the holes until you step in one. Also, opportunities come up during the project that will change the scope – ex: you need five checkout methods instead of four. Make sure the scope covers what happens when these situations arise because they will happen.

Cost

➲ What will the costs be? Are they fixed or T&M (time & materials)? When will payments be made? Will payments be made by established deadlines or by work completed? All payments should be detailed and tied to a milestone.

Payment schedule

➲ Develop a detailed payment plan. The plan should be based on either completion points/milestones, or dates. As long as all parties involved understand when the payments are to be made, the project should stay on schedule and finish on time.

Software Client detailed scope of work enabled the project to be successfully implemented. A correctly crafted scope reduces "scope creep issues," a term used to describe work going on outside the scope. Many projects fail when the scope is not detailed enough; detail increases the likelihood of a successful project.

2.5 TECHNOLOGY OPTIONS

Purpose:

Understand technology and make the right choices.

Key Points: **In this section you will learn how to evaluate your technology options. Your decision will be based on budget limitations, long term goals, access to resources, and internal & external talent.**

The spectrum of the World Wide Web is so large and crowded that many technological choices become lost in the shuffle. According to a study at the University of California, Irvine, a common set of selection metrics was not possible when making a choice on technology implementation because terms did not have the same meanings for different technologies. In actuality, the formal procedures were interwoven with sense-making activities so that technologies could be understood, compared, and a decision consensus could be reached. This means you are really at the mercy of the team you work with (whether internal or an outsourced vendor).

You may hear what the "experts" say about which technology to use. Hundreds of articles cover the positives and negatives of every possible technology option. Development firms have their own biases because of their own approach or internal strengths. It becomes very difficult to make the right choice for your project. With so many options how do you make the right choices? Who do you trust?

CUSTOM NEEDS

Software Client approached XploreNet requiring a Microsoft .Net programming platform. They had identified the .NET platform as the best option; however, they needed XploreNet for confirmation of that choice. The answer to making the right technology choice always comes down to you and your firm. What works best for a firm like our Software Client may not work for your organization. In fact, there are few workable cookie cutter solutions. You will need to work with your entire IT and executive staff to make sure that the technology you implement for the Web fits the current platforms, staff objectives, and future plans.

CATEGORIES

Web-related technology falls into a few categories including:

➲ Graphic Development. Tools that assist with graphic creation, picture resizing, format conversion and file/graphics optimization fall into the graphic development category. Photoshop is the most commonly used graphic development program. This helps ensure the graphics, pictures, and overall design fit the page and match the brand.

➲ HTML editors. A second category is an HTML editor or WYSIWYG (an acronym for "what you see is what you get"). Both are tools used to create the HTML code of the Web pages. The most common program choices are Dreamweaver and FrontPage. Many technology savvy individuals and firms use Homesite (now called Cold Fusion Studio) which is a text editor that color codes the test.

➲ Programming tools. The third category is programming tools which are designed for languages like Active Server Pages (ASP), .NET programming, java server pages (JSP) and PHP. These languages are used to build the database functionality in most sites and display static (HTML based) content/pages, from processing a submission form to accessing a database.

➲ Databases. Lastly, databases are the tools that are used to make a site truly dynamic and ever changing. They send updated information to the web site to be displayed and examples include items on sale in an ecommerce site to the current amount owed on your credit card. SQL, Oracle, MS Access, and MySQL are the most commonly used.

One way to ensure you make the right choice is to list the different categories as shown below:

Budget levels	Database Options	Programming languages	Business objectives
What can you afford and what makes the most financial sense?	Based on your budget and specific objectives which database will work? This can depend on the size of the database and expected load (how many users will be using it)	Based on the skill sets you have in-house or the firm you choose to work with, which will achieve your desired results? Also, what type of platform the site will reside on and what external systems it will interface with?	Which objectives require which technology options?
Key areas: Budget Restraints Short/Long term goals	Key areas: Hosting options Software fees	Key areas: Beginner vs. advanced skills Complex vs. simple programming	Key areas: Sales and marketing Internal communication

The diagram above shows four columns: budget levels, database options, programming languages, and business objectives. For more detail, add HTML editing and graphic development software columns. By answering the questions listed in each column, you gain a strong feel for the best choice you can make.

CHOOSING THE RIGHT TECHNOLOGY

If you have the internal resources to identify which technology is best for your firm, use them. If you lack those resources, bring in an outside expert. Software Client worked with XploreNet to make sure that the Microsoft .Net platform would work well with their goals. They required an outside professional opinion.

Here are some tips for choosing the right technology:

- ⮑ Build from the strategy. Your technology choice should enhance the strategy you have begun to develop.
- ⮑ Research. A little research goes a long way. Look at how other firms are improving marketing and sales through new Web technologies.
- ⮑ One step at a time. It is not necessary to implement all the options by tomorrow. Develop a phased approach.
- ⮑ Establish the budget. Set aside time & money, use part of the budget for experimentation (tools that you may or may not end up using).
- ⮑ Focus on information and plans. You can increase your own power with information and connect with more customers through a well thought out plan.
- ⮑ Utilize leverage. Use technology that helps you do more, with less personnel and financial resources.
- ⮑ Keep it simple. Pick the languages, the tools, and the equipment that are easy for your staff to work with.
- ⮑ Use outside consultants. Get help from external resources for technical or process issues.
- ⮑ Start today. The success of your firm may be defined more by your technological choices than any other factor.

Software Client was able to confirm their technology decisions because they had an understanding of all the strategic, tactical and operational factors related to their business objectives and strategy (see Module 1). They did not guess at the best option; they worked with a professional firm to identify key areas and anticipate potential problems.

2.6 CONTENT

Purpose: **Understand how content will be gathered or created and to whom the responsibility falls for each portion of the content.**

Key Points: **In this section we break down the critical step of Web content. A checklist of your content is required. The checklist will help everyone involved understand how to organize the content and deliver it in a timely manner.**

After you've completed your analysis and have a good understanding of the user's behavior, the scope of work, and the technology options, you need to evaluate the content for your Web project. For our purposes, content includes text, graphics, pictures, and specific tools. Content is what makes a new site effective or ineffective. The *toughest* problem to overcome during any project is gathering content. The project gets held up and key steps are sacrificed because the right content is not delivered to the right people at the right time.

Quality content increases repeat visits from users and raises their level of trust. Content on the Web should be simple and concise. If you do not know what your visitors want, ask. A clear understanding of what your content should say (commonly called "copy"), and how pictures and graphics will enhance the presentation will increase the success rate of your redesign. And most importantly, make sure deadlines are realistic and that everyone on your team understands their role in obtaining the content.

IDENTIFYING PROPER CONTENT

Gathering and creating content can be one of the biggest challenges of a redesign. Software Client had many user groups to which to match content, and the new content had to be in-depth and user-friendly. They knew that just utilizing the existing content and simply giving it a face-lift would be an ineffective solution. Software Client content had to be straightforward, concise, and relevant. If the content for your Web project is unclear, *don't* start your project. Software Client knew where some of the new content would come from, but they had to match the user's goals to specific content areas. This required user feedback, and careful consideration of the responses. By taking the proper steps, you will know what type of content should be created. Use the following guideline:

Step 1:   Create a content outline.
Step 2:   Assemble the proper materials.
Step 3:   Brainstorm with your staff on content categories and areas (for example – a customer service resource center).
Step 4:   Think in terms of quick sound bytes.
Step 5:   Prioritize ideas.
Step 6:   Create a task list.
Step 7:   Implement the proper content.

CONTENT SOURCES

If you lack key content research services, search for sites that provide content related to your industry or topics. These are often referred to as syndicated content streams. This type of content is provided by other sites, for use on your site. You can simply place a few lines of code and content is delivered. The linked service provides the content to your site automatically with no need for maintaining the code. The content provider either charges you a fee for using the content, or they receive traffic from your site via links placed directly in the content. Interestalert.com will pay you for adding a news feed to your site. There are many free services and budget sensitive options. For example, www.7am.com provides specific news content, while www.1afm.com provides free content. Another great resource is www. yellowbrix.com, providing industry specific news and updates.

CONTENT INVENTORY

Software Client identified what content would be important and created a content inventory. They utilized an excel spreadsheet (a database works well, too) to keep it organized. They created a check-list of the existing content, taking a hard look at each content area, and asked:

- ➲ "What's missing?"
- ➲ "What still needs to be created?"
- ➲ "How long will it take to create?"
- ➲ "Who needs to create it?"

They were realistic about their time frames and knew that if content piece C was not in place then D would have to wait. A good outsourced partner will help you understand this process and give it to you straight.

WEB COPY

One of the most difficult yet important areas of content is the copy. It can be costly to create copy in time, finances, and resources. If you need to create content, designate a qualified internal staff member to do the writing or outsource the copywriting to a qualified copywriter. Poorly written copy reflects on your company's professionalism. Don't give your users any reason to doubt you. After designating the copy provider, establish a schedule to keep all personnel on track.

Writing copy for the Web is different than other marketing channels. There are several differences, but one thing remains constant – focus on the user. What will be important to them? What will improve their chances of making a purchase decision? What type of copy will bring them back for another visit?

According to one expert, you should only write 50 percent or less of the text you would have used in hard copy. Only give your user the most pertinent information. However, be detailed when necessary, and make sure your potential customers understand the products and/or services you're offering. The Web is no place to display the full content of a book. Most users go to your site for quick, specific information - not the history of the product. Providing quick bullet lists whenever possible keeps the eye moving yet highlights key points in your copy. Keep the following tips in mind for the text of your Web project:

➲ Break pages up into short blocks. Many users become frustrated when they must page down continuously. Under no circumstances should you allow your users to become frustrated. Use different sized fonts to show important comments, and hypertext to break up pages and split long sections or tedious information. Most pages should have similar formats regarding paragraph set up, font type and size, and overall text look and feel. If customers think they've linked to a different site, then they may not come back. Web pages should have a consistent amount of text on each page.

➲ Double check, triple check, and have every resource at your fingertips look at the site for misspellings and errors. One secret to decreasing misspellings is to copy all your Web text into a Word document and hit the spell check button. Misspellings and errors cause your visitors to lose confidence in your company; proper grammar is also critical. It can't be emphasized enough that your Web copy must be written by a qualified professional.

➲ Focus the words on customers and how your service/product helps them. Stay away from "I" and "us," and concentrate on "you" and "we". Keep text positive and forward thinking, but do not exaggerate. Offer convincing words that list the benefits you offer to users and your company's enthusiasm for their business.

GRAPHICS AND PICTURES

The second category of content is graphics and pictures. First, utilize internal pictures/graphics only if they are of high quality and/or professionally done. New marketing material typically contains the right pictures and graphics. Make sure the ones you use for the Web match the offline marketing and branding packages (same feel, colors, look, consistency, etc.). Second, there are many types of libraries and online tools to capture the appropriate graphics/pictures to match your project. Perform a search on the Web with the key words "free Web graphics". You'll find plenty of libraries offering free or modestly priced graphics and pictures of the right quality. Also, invest in a subscription to an online library to purchase pictures/graphics as you need them (www.corbis.com is a good example).

Software Client realized through the blue printing process that quality content is important to visitors returning to their site and in creating the perception of expertise. There is a fine line between too little content and too much. Understanding your content needs during the Blue Print Phase will confirm whether the project schedule and timelines are realistic.

2.7 INITIAL DESIGNS

Purpose: **Establish initial designs based on the internal team's perceptions and ideas.**

Key Points: **Initial mock-ups will cover what the internal staff and Web vendor think are important. It is during this process that the initial usability testing begins. The final product may look nothing like these initial ideas; it is the process itself that is important.**

As mentioned in the beginning of this book, redesigning a Web project can be frustrating. If the design (defined as the "look and feel" and for our purposes in this section, the "navigation") is poor, then business objectives may not be met. The design you present to the external world creates user perception. Start with your internal team, but rely heavily on outside opinions especially your customer's feedback.

XploreNet has built over 650 Web sites and applications. Some tried and true theories have become very apparent through the years. When you are blue printing the project, you must come up with some initial design thoughts. Do not wait until you are building the pages. Start talking mock-ups from the beginning.

INITIAL DESIGNS

Creating a design for the Web is very different than designing a brochure or printed document. XploreNet worked with the Software Client Web committee to come up with some initial design thoughts. This included the main page design, also known as the graphic interface. XploreNet worked with the Software Client to make sure the new design gave the user control over the experience. Modern Web users expect a level of design sophistication from all Web pages. The goal is to provide for the needs of the target users, adapting the technology to the user's expectations. Never frustrate users by designing roadblocks. The main page's purpose is to aid the user in performing a variety of tasks.

Initial design thoughts are based on the home page or main page of the site or application. The main page acts as a common sense entry point. However, keep in mind that some users will enter into the site at sub-page levels. For example, our Software Client would have visitors coming from search engines, banner ads on industry sites, and seminar material. These outlets would, in many cases, offer different landing pages other than the main page.

The main page is roughly 28 square inches. The top of the page comprises the key real estate on the main page. Most users will immediately look at this part of the page, especially since most users view the page In a 1280 by 720 monitor setting. Therefore, positioning is critical. Similar to a newspaper, stories at the top of the page are read more than those on the bottom. Utilize the top of your main page for the navigation bar and key messages.

Initial design strategies for the main page fall into three main categories:

Channel-based main pages

➲ This main page strategy is typically utilized by firms with large sites offering so much information to several different user groups that putting all the appropriate links and content on the main page would cause the page to be overwhelming to all visitors. Furthermore, users visit these types of sites for specific information or tasks. If your site is this large, then it would likely make sense to divide it into different categories and offer more details in specific sections. For example, you might have three key divisions in a company. The menu options might be Asia, North American, and South America or students, teachers and non-profit organizations.

Menu-driven main pages

➲ In the early days of the Internet, and even now, text based links dominate main pages. This type of design is dominated by plain lists of text based HTML hyperlinks. More complicated designs combine blocks of text links and graphic links. Typically text links are much easier to edit while graphic links are more space efficient.

News or information main pages

➲ This type of design strategy is used by non-profit organizations that sell nothing or news organizations (USA Today, CNN, etc.). These firms utilize their main page to make announcements on what's new or the day's main news stories. This is information that is updated daily or hourly, thus the site is more likely to experience repeat visitors. Most firms utilize their main page for event updates or late-breaking news. In order to utilize this strategy to its full effect, make sure the location of the news or updated information stays consistent. If you make changes or revisions often, keep this information in the same place.

Software Client chose to work with a menu-driven design. The company came to this decision because after going through the Analysis Phase and at the beginning of the Blue Print Phase, it was discovered that their calls to action were more appropriate for this type of strategy. As you start to prepare the initial mock-ups, fall back on your user profile. It's difficult to design for all audiences or unknown users whose preferences you know nothing about. Software Client matched the users and their scenarios to the design, asking the question, "Would a targeted user, attempting to find specific information or complete our call to action, be hindered by the main page design?" Performing usability testing and getting feedback is the best method by which to determine if your main page gets your message across accordingly. There are several critical areas to think about:

Match the main purpose/goals to the design
➪ When you begin a project, you must have your priorities already established. What are the most common objectives? You defined them earlier in the goals portion of Module 1. Now it's time to connect the dots.

Focus on the main page or main user interface section
➪ This is the foundation of your design – every thing should be built around this primary component. The navigation system, overall look and feel, brand objectives, and calls to action should now be considered. From the main page, you can identify subpages and required content.

Complete mock-ups of all key user pages
➪ Create mock-ups of all pages where a user is given a call to action such as purchase a product, inquire about a service, log in to an account, etc. These pages are simply Photoshop files with minimal HTML or functionality. If done properly, the mock-ups can help reduce time spent rebuilding pages. Do not build out any HTML until the design is agreed upon.

TESTING YOUR IDEAS

Getting opinions on your mock-ups can be a difficult process. Web pages can be like art, one person's treasure is another person's trash. If you have the budget, do a full usability test (which includes five different categories and a set process). If not, perform a simple one. Choose three to four designs that fit most of the goals from Phase 1. Locate four to five users from the defined user groups and test (The next section covers usability testing in detail). Create two to four mock-ups for each test page. Ask questions, have the testers work with the navigation, and establish your calls to action. For example, how would you (the user) contact us? How would you purchase services from us? What content looks appealing or causes curiosity?

AVOID SPLASH SCREENS

A quick note on splash screens or intro pages: while attractive, they should typically be avoided. They serve no good purpose and can annoy the first-time user, and especially the repeat user. They are often built in Flash, which can take a long time to load or may not load correctly.

Software Client realized that the initial designs would set the foundation for the rest of their site. At this point, you will make your design decisions based on past user feedback and internal ideas and goals. Although the overall "look and feel" will make the first impression, the organization of the menu and key information will either help the user in the next step or drive the user away. After the usability testing, they knew whether they had hit the target or needed to go back to the drawing board. Finally, search engines still struggle to correctly index splash pages

2.8 USABILITY TESTING DEFINED

Purpose: **Understand how your visitors will use the site or Web application via usability testing.**

Key Points: **You will use your initial design mock-ups to test your ideas. Users will either confirm your theories or show you where you missed the mark. It is critical to be impartial in this phase. For example, you may really love the color orange, but the user doesn't think it matches your objectives. Listen to the feedback and implement what works best for the _user_.**

When you look at your favorite Web sites, what about them appeals to you most? What do your Web site's visitors like about yours? It's not uncommon for your different customer/prospect segments to behave differently. Each segment may need different types of information and be drawn to specific parts of your Web site. How do you know your Web site is delivering the information customers and prospects need to buy from you, versus your competition?

For nearly as long as the Internet has become a popular medium, XploreNet has designed Web sites and Web-based technology solutions. And from this vast and diverse experience of serving clients in multiple industries, we've learned the best way to know what your customers and prospects want, **is to *ask* them**.

LACK OF USABILITY TESTING

If you've ever visited a Web site or online tool and felt frustrated, lost or downright disappointed, it is often because of one key element: when the site was built, no usability testing was completed. Software Client current Web site did not have any usability testing conducted before it was launched. They didn't ask their most important group (site visitors) how they felt about the site and other questions, like: Does the navigation make sense? Does the user understand the calls to action? The easiest way to lose a visitor is poor usability.

USABILITY TESTING PROCESS

Understand Web usability by performing the Grandma test. Imagine your Grandma sitting in front of her PC, accomplishing a list of tasks. As she works through the tasks, she provides feedback, rates the difficulty, and the observer makes notes. If a Grandma (in this case, stereotypical of the most basic Web user) can navigate and accomplish the calls to action, then the majority of your audience should be able to do so as well.

XploreNet conducted usability testing with 10 participants for the Software Client, utilizing the mock-ups created in the initial design sessions. We performed the usability testing procedure outlined in Module 5 in the Web Plan.

Figure 2.8: Usability Test Process Flow

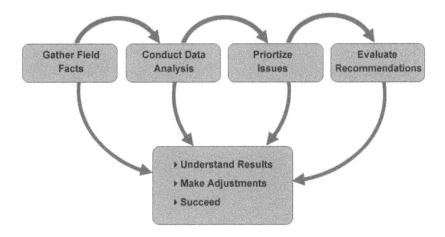

IDENTIFYING PARTICIPANTS

You will need to identify a proper group of participants (beginner to advanced users, as well as a diversity of demographics including men, women, younger to more mature, etc.). The goal is to test users from the site's target audience -- complete tasks typical of those they would do "in everyday, real-world scenarios". For example, the Software Client wanted their users to know how to shop online, how to find the right contact for enterprise software purchases, and how to sign up for the e-learning courses they offer. Your testing procedures might ask the user to input user names/passwords, find contact information, and/or purchase or donate something.

All participants have different ability levels – beginner, intermediate, and advanced. Furthermore, make sure participants are asked consistent questions and tasks in the same sequence. The tasks should be divided up into broad-brush tests and priority questions covering a few important elements or calls to action.

TEST QUESTIONS

You should identify 15 to 25 key questions, including a three to five question pre-test designed to gather background information. The pre-test typically includes demographic information such as level of Internet/computer experience, age, company role, gender, type of computer, browser they prefer/use, connection speed to the Internet, etc. During the test, ask participants to accomplish specific tasks (i.e. sign up for something, contact a specific office, etc.), and make sure you understand how the user uses the site. Tabulate the results by setting up the data in a database format or as a simple Excel spreadsheet. Finally, make the appropriate recommendations for the proper design and development based on the data results.

TESTING METHODS

Once you know what questions to ask, decide on the most appropriate testing method. The most common testing methods include:

- An email survey sent to a specified group - typically 100 to 500 participants.
- Perform testing via telephone interviews while the subjects view mock-ups and navigation on their computer.
- Perform testing in a laboratory, where you may record both audio and video results.

All interviews should last between 25 minutes and an hour. For this client, telephone interviews made the most sense. XploreNet contacted the identified participants and walked them through the process, completing a pretest questionnaire before beginning. The results of the usability testing are not intended to be the sole decision criteria of the final design. Instead, they are a guide. Combine the data with the other details you uncovered in the Analysis Phase. Software Client considered all factors, including the initial designs, usability test results, and overall company objectives to determine which design would work best for their targeted users.

2.9 TECHNICAL PERSONNEL

Purpose: Understand which technical personnel should be involved and the proper time to bring them on board.

Key Points: In order to know when to bring in the technical staff, you must know both their strengths and weaknesses. The more complex the project is, the earlier the technical staff should be involved. Make sure you have set specific goals for your technical staff.

For a web project to succeed, you need to understand the important role the technical personnel will play in the success of the project. Timing and the process you use to engage them is a critical part of your web plan. If you've ever participated on a web project with a technical person, then you might have experienced one of a few things. Some intelligent IT individuals are easy to work with, get along well with others and implement some great ideas while others can be abrasive and challenging

WHEN TO GET THE GURUS INVOLVED

You should get your technical gurus involved as soon as possible on most Web-related projects. A technical guru typically approaches the project strictly from the technical side. Often, they do not have a clear understanding, nor are they concerned what occurs in the user interface or the marketing aspects of your Web project. Remain aware that technical gurus rarely think beyond their own technical scope. Therefore, keep your technical gurus in check.

DEALING WITH THE TECHNICAL PERSONALITY

Technical personalities tend to focus on how something will work, observing the proper code and the functional process. They may often overlook things like calls to action, sales process, and content relevance. However, their input is incredibly important. They are often a very effective sounding board or idea testing tool. You may have great ideas on the front end or user side, but they may not be practical when it comes to your resources, budget requirements, timelines, or functional requirements. The technical guru can also help to keep you in check.

Software Client made certain their technical personnel involved on the project were given the opportunity to voice their opinions and concerns. Great ideas can be killed by not letting the technical gurus provide feedback and make suggestions. This should include both your internal staff and your outsourced partner. Make them part of the entire solution. Help them to stay on track, listing priorities and technology options, covering timetables and platform options. The more you inform the technical gurus about specific project goals and timelines, the better the interaction of the technical team. In the end, this results in a successful project.

2.10 FUTURE PLANS

Purpose: **Understand how your Web project will evolve over the next 12, 24, 36 months.**

Key Points: **Future Web plans are not like future business plans. Web plans are shorter and must take into consideration ever-evolving, new technologies. However, continue to match business and marketing objectives to your future Web plans.**

Planning for the future when covering any business function is difficult, but the process becomes even more difficult when the web is involved. Since we don't have a crystal ball, we have to go with established best practices and take a shot at making lots of assumptions. Some will be correct and many (maybe even most) will probably be wrong. At XploreNet, we focus on real business solutions combined with future implications. There are flying cars but no highways for them which you must keep in mind when planning for the future web.

In addition, the latest buzz phrase is Web 2.0. Tim O'Reilly, founder of O'Reilly Media, suggests that the idea of Web 2.0 "relates to the transition of web sites from isolated information silos to interlinked computer platforms that function like locally-available software in the perception of the user". Basically Web 2.0 is referring to a web browser/server driven connectivity. Whatever lives on the hard drive of your PC will have less value than your connection to the Internet. You have to anticipate some of the impact(s) that Web 2.0 will have on your business operations.

One thing is certain with the future – it will be technology/web driven and lots of changes will come (as they always have). Start planning for it now. Compare the web to a house with movable walls, floors and ceilings. The plumbing and electrical are components like Web 2.0 that cannot be moved around once they are installed. Once you make this analogy, you should easily see the impact that Web 2.0 will have. It will shortly become very important to be careful when choosing how your business will connect to the internet. Outright ignorance could cost your firm thousands of dollars in redesign if you don't have all the facts to make the right decisions.

WEB, E-MAIL, AND SERVERS BECOMING ONE

The difference between the web, e-mail and internal networks is graying more with each new day. Now there are blogs, podcasts, and file sharing. All the tools are becoming one. Make sure you keep your eye on these changes, noticing how you can take advantage of the opportunities that will soon appear.

THE IMPACT OF SOCIAL NETWORKING

The social network has become the latest tech phenomenon, connecting all the people of the world (not just in the US). Your web strategies will change based on the fact that social networks are becoming more powerful through the sheer number of users that utilize them. The communication tools include Instant Messaging (IM), cell phone, iPods, text messages, and e-mail. The growth of these social networks means that your business must communicate and market to your customers and prospects differently tomorrow than you do today.

TECHNOLOGY CYCLES

Technology is ever-changing and doing so rapidly. The latest statistics provide a 12 to 18 month cycle, meaning that new technology is coming every 12 to 18 months and in some cases every six months. This uncertainty made it difficult for the Software Client to estimate what changes would need to be implemented in the future. However, they understood the importance of establishing how to work with new technology, and the effects on the company's business objectives and rules.

DEVELOPING FUTURE PLANS

Even though it is difficult to predict, you must still plan for the future. How do you do this? First, understand that any Web-related project, whether it is a new Web site or new Web application, is a continually evolving process. Software Client new Web tools were built on the Microsoft .Net platform, which is recommended when future enhancements and growth are anticipated. As your business changes, so will changes be made in the technology required to operate it. Therefore, establish both strategies and tactics. Tactics should define how your new Web strategies will be implemented. Software Client tactics included future personnel needs, the number of reoccurring small projects, add-ons, and special future SalesLogix applications.

When it comes to the Web, you have a large advantage – it is much easier to change than most aspects of a business. Although technology projects can have steep price tags, these are comparable to changing your location. Programs can be rewritten in far less time than it took in the past. New tools can be added on, often in only weeks or months, rather than years.

Future Web plans take into consideration your company's future business environment, staff needs, customer service, production processes, and sales & marketing systems. When you look into the future, try to understand how the new technology will help or hinder these key areas. Create an evolving development process that fits your current business model and shapes your future business plans.

COMPLETION OF MODULE 2

You have now completed the Blue Print Phase of your Web plan. Refer to Module 5 for the outline and details of a properly constructed Web plan. The Blue Print Phase is critical to the success of your Web strategies and the projects that enable those strategies. After completing modules one and two, the foundation of the Web plan is completed. Our experience has taught us that taking short cuts during the first two modules may seem easy and make sense but can cause frustration and major losses of both time and money down the road. Spend some extra time in these phases and your ROI will go up significantly. At this point, the Software Client knew who, what, how, and when their Web strategies would be fulfilled. The next step of completing the actual construction of the new Web project can now be successfully and efficiently completed.

MODULE 3

CONSTRUCT

3.0 Construct

This module discusses the next phase of understanding how your Web project fits into your Web plan. You will quickly learn how to match the deliverables to an end product. This is where you build out the Web project. In other words, if you were putting up a building, you would have the land bought and the blue prints completed. Utilizing the information obtained in Phases 1 and 2, you have the right information to begin and complete the construction process successfully.

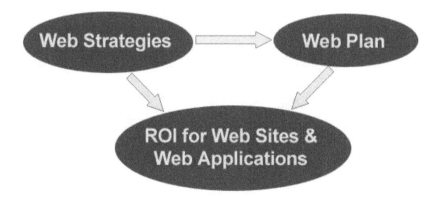

Consider the similarities between assembling a building and a Web project. The final product is only as good as the engineering strategies and the blue print. In this module, we cover important areas like architecture and navigation, project management processes, project management tools, the communication process, and quality control.

Most books and Web firms cover the construction phase of a Web site strictly from the technical view point. Furthermore, they typically recommend technology that they know and not what might make sense or be best for *your* strategies. However, the construction phase of any Web project is the implementation of the strategies (Analysis Phase) and the design issues (Blue Print Phase). This is where the written details become *real*, enabling your team and the users to see and use the end product. If the first two modules were covered correctly, the construction phase is simplified.

3.1 ARCHITECTURE & NAVIGATION

Purpose: **Understand how the project blue print will be implemented.**

Key Points: **In this section we cover the comparison of building a Web site or application to putting up a large high-rise building. You will make the connection and understand why issues can become complex very quickly. You will understand the importance of having detailed architecture and simple navigation.**

The City of Denver has recently finished construction on a new convention center and now several hotels adjacent to the center are rising from the ground. Each day people see new levels completed, new equipment put in place, and different personnel with different roles. By understanding how a large building is constructed, you'll have a better feel for how a complex Web project is built.

ARCHITECTURE & NAVIGATION

You may be asking yourself how erecting a building relates to a Web project. The answer may surprise you. The key is your project's architecture and navigation. Instead of pouring concrete, moving heavy objects in place, putting up walls, and installing electrical and mechanical systems, you're implementing a blue print for success on the Web. It is amazing how architects create blue prints that show construction superintendents and foremen all the right parts, to fit at all the right times, in all the right places. There is a remarkable synergy between buildings and Web sites, and the same elements and processes between architecture and Web development. You can garner some great "tricks of the trade" from the architectural design process that can be applied to building a new Web project.

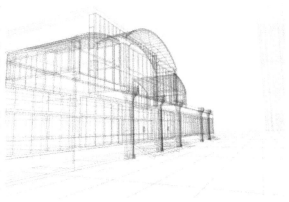

The challenge with most Web sites is that the developer(s) did not look at the site as an architect would look at a building. In fact, many sites would be condemned if they were buildings. If you could provide all the information on one page, navigation wouldn't matter. The modern Web makes this almost impossible, so you need a quality navigational structure. Sites and tools that require you to click four, five, and six times into the site or have to click back six levels to get to the main page have completely missed the target. If you had a building that required you to figure out how to get to the fifth floor with no signs or elevators, your visitors would be in trouble. Many sites do this to the user and this hurts business.

THE DIFFERENCE BETWEEN ARCHITECTURE AND NAVIGATION

In order to build a great site or Web project with simple navigation, you must understand all the areas we've covered in Phases 1 and 2, as well as understand the difference between architecture and navigation. Although the two work hand in hand, they are unique. Navigation on the Web is defined as the user moving from one page to another within the site; architecture is defined as the design of those pages (both static and dynamic) and the connection between them. A good way to think of this difference is that architecture is the arrangement and make-up of the content, while navigation involves the devices that move the user from one page to another.

USER PERCEPTION OF NAVIGATION

The user's ability to navigate should be as straightforward and simple as possible. Keep in mind that users will not memorize your navigational structure. "At IBM and at Sun, we studied how people read on the Web. What we discovered is they don't read! They scan.," says Jakob Nielsen, a distinguished former Sun Microsystems engineer. The user is scanning the pages for key words, phrases, and headings. Finally make it simple for the user to follow your information in a process – step A, step B, step C, etc.

Some navigation structures are involved, pushing users to certain sections of the site. The problem is that users won't take the time to learn the logic of your navigation. Users want the right information to lead to the right call to action, right now. If the navigation is understated, the user will get lost; if the navigation is overwhelming, they'll simply leave (close the browser).

You must create the navigation with a logical process in mind because a high percentage of the visitors to your site will be repeat visitors. Once they've been to your site several times, they'll know where to find what they need. However, the navigation should be very obvious to the new visitor as well. Simple navigation enables the new visitor to find what they need quickly.

EDUCATING THE USER

No matter how visitors get to your site or Web application, chances are they are going to be unfamiliar with it. If they get to the site via a search engine, then the process gets even more difficult. They may be dumped into a section other than the main page - even a back page that is not viewed regularly. Therefore, it is important to make sure that the main menu and key links are apparent on each page.

Here are some strategies to help your visitors navigate through your site when they come from a search engine.

➲ Craft a clear message about the purpose of the site.
➲ Usability is your key.
➲ Define useful and simple navigation.
➲ Include critical information that the user expects.
➲ Develop beneficial content.

GREAT NAVIGATION

Great navigation is about simplicity and common sense. How do you create an environment that facilitates users to navigate your Web project correctly? One important element of navigation is called top-level navigation, and consists of the main page of the application or main page of the site. Pay close attention to the main navigation structure or main menu choices. As most people know, this is either the menu across the top or down the left hand side of the page. Now, what makes sense to the user? Refer back to the Usability module for more details.

Architecture and navigation work hand in hand on the Web, assisting the user to get what they want. That is the real measure of successful architecture and navigation. Did the user find what he was searching for? Simple navigation works and typically costs less to build and maintain. Here are a few quick hints to help the process:

- ➲ Enhance the site with interaction (tell them to do something).
- ➲ Make navigation simple and intuitive.
- ➲ Use external links and implement a process to update those links frequently.
- ➲ If possible, provide a search function for the site's content.
- ➲ Avoid the need to include a scroll bar whenever possible.

3.2 PROJECT MANAGEMENT PROCESS

Purpose: **Evaluate the proper process to implement your scope of work.**

Key Points: **In this section we'll cover a potential project process. If you follow an established process that incorporates key project management tools and requirements, you're more likely to implement a successful solution.**

Project management is one of the most important components of completing a new Web project and is often the component most overlooked. A full life cycle project management process is the vital link that holds every development project together. In fact, you must have buy-in from your top executives and a plan in place to accomplish the following: defined project plan, schedule, resources, budget, risks, and scope. The key elements to a successful project are validation, organization, initiation, management, and completion. When you approach a redesign or the development of a complex tool for the Web, you must understand how the process will work. A well-organized, simple process will increase your success rate, while a lack of one will very likely result in failure.

PROJECT MANAGEMENT OVERVIEW

The field of project management has been studied in great detail, yet many of the processes are difficult for people to grasp and articulate. Since 2000, the number of studies related to project management has increased significantly. The results are still mixed. According to CIO Insight, CIOs are knee-deep in IT projects these days. But when it comes to managing these initiatives, companies often lack discipline, thanks to half-hearted involvement and insufficient follow-up. Here are some statistics provided by CIO Insight:

➲ 53 percent of CIOs say their IT project prioritization is politically driven.

➲ 40 percent of CIOs use a portfolio management approach to IT projects.

➲ 66 percent of IT projects during the past 12 months came in at or below budget.

➲ 13 percent of IT projects failed to meet the goals of IT and business management.

THE PROCESS

Many firms forget that completing a usable, successful site requires a process. The firm or rather, their clients are often disappointed with the end result, blaming the mistakes on poor design, when the reality is they did not utilize the right process. Their project management process was incomplete or non-existent, causing delays, misunderstandings, and frustration.

The project management process is not about completing an initial planning meeting and then going to work. It's about daily or weekly updates, identifying what works and what doesn't. You'll need to initially validate the new project (covered in Module 1). Get organized and then relax because you now know how all the pieces will fit together. Get started and never look back. Analysis can become paralysis very quickly. Furthermore, commit to ongoing management whether the project takes six weeks or one year, or more. The best part is always the completion point. You'll be done before you know it. In addition, using the scope of work as the blue print, draw a flowchart for the process. If you're working with a development company, they should have a process for you to utilize. If you're on your own, refer to the XploreNet Web plan at the end of this book or email byoung@ xplorenet.com to receive information on the proper process.

Figure 3.2: Project Management Process Overview

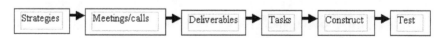

Understanding the details of this process will allow the delivery of successful results. There are many key issues including planning, implementation and quality control. There are also challenges in every step, which will need to be overcome. You may have several people involved with different goals and agendas. The challenge of understanding who does what and when can be overwhelming. Several key areas are involved the process:

- ➲ Identify who the lead person or persons are.
- ➲ Define the method for completing each step and getting to the next one.
- ➲ Define who is responsible for what.
- ➲ Establish realistic timelines.
- ➲ List key questions and answers.
- ➲ Create a task check-list from the deliverables, using the following tips:
- ➲ Find somewhere quiet, with no interruptions, to complete the checklist.
- ➲ Create checklists at the start of the project, near the middle, and at the end.
- ➲ Print out a rough draft list, adding notes as you go along.
- ➲ Review each task on the list, go away, come back, and either confirm or delete it.
- ➲ Identify which tasks will cause the biggest challenges.
- ➲ Make sure the entire team understands all tasks.

PROJECT MANAGER

A good project manager is critical for a successful project. This includes both the person in charge of your Web project at your firm and at your outsourced vendor. The project manager has several goals, but one focus: finish the project on time and within budget. He or she needs to be able to handle administrative tasks, the people involved, and the details of the Web plan – all simultaneously. This means the project manager's main skills are to communicate effectively and to multitask. He or she also must be accountable for the entire project. Accountability on most projects is often overlooked. This means defining who is responsible, and for what. Make sure you have it clearly detailed in the plan as to who is not only responsible for what task, but who is responsible for the strategic initiatives and tactical implementation.

TIMELINES

Make sure timelines are clearly defined and understood by the entire team. Ask the key question, "Are our timelines realistic?" Have a back up plan for challenges that arise during the project. Change and challenges always come up, so don't get stressed when they appear – be prepared for them ahead of time. Have a detailed process, often called change/challenge management for handling the scope and technology changes. New ideas and better methods can emerge during development so don't kill them if they deviate from the original blue print. Instead, have a process to facilitate them.

A correct project management process will help all the players understand their roles. Furthermore, the process will help monitor project activities and deliverables against the plan. It should help with tracking challenges, communicating progress, and identifying the successes and failures learned for the next project.

3.3 PROJECT MANAGEMENT TOOLS

Purpose: **Evaluate the proper project tools to implement the scope of work.**

Key Points: **Understand the types of tools and view an example of the proper project management tool. You should know what categories to cover and how the tool(s) help both the internal team and the outsourced vendor.**

Once you have a properly aligned project process, you must have the tools to accomplish your most important initiatives at the right time. There are a million choices when it comes to project management tools. We do not make specific recommendations, but rather provide some general guidelines and helpful hints of what the chosen tool should contain, at the minimum level. As you read through the options, you'll begin to see how matching your needs to the tools makes the choice for your Web project clear.

Utilizing a tool that fits your needs provides many benefits. If you've used a project management tool in the past, you'll see what might have been missing and if you've never used a Web-based tool, your life is about to get a little easier.

PROJECT MANAGEMENT TOOLS

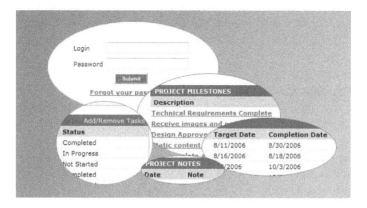

Project management tools come in many forms and you may even have an internal project management tool in existence. Here are a few "best practices" for a project management tool. Make sure any system you use has at least the following components:

Security
➲ The tool should be secure and only allow access via a user name and password. It is most productive to have a unique username for each member of the project.

Web-based, 24/7 access
➲ The tool should be on the Web, accessible 24 hours a day, seven days a week and not require any special software downloads or plug ins.

Scope of Work
➲ The tool should have a link to the scope of work to enable all involved in the project to see how the work has been defined.

Task List & Task Tracking
➲ The tool should enable both you and your outsourced vendor to update and add to the task list and have a built-in task tracking system. The system should monitor tasks assigned to individual team members and alert both internal and external team members when a task has been missed or changed. This prevents a small problem from becoming a large one.

Automated Emailing
➲ The tool should have the capability of an automated "elective emailing system" that performs peripheral tasks such as sending appropriate emails to project team members, whenever required. This ensures you are free to concentrate on the core project management functions.

Upload Area
➲ This area can be utilized by either the internal or external personnel to FTP content, text, graphics, links, etc. to the project management tool.

<u>Document Sharing</u>

➲ Make sure all team members can upload files and documents related to the project. The files and documents section should contain all the project related files and documents that personnel may have posted for the review process. Sharepoint has become a favorite tool for managing documents.

<u>Discussion Board/Communication Center</u>

➲ Utilizing a threaded discussion area improves communication between internal team members and the outsourced vendor exponentially; it also aims to eliminate communication gaps that exist within teams and in the correspondence with clients and other external agencies. One way to accomplish this is to enable team members to post their views and questions, to be answered or commented upon by other members. You may see the discussions for all tasks or see them listed separately.

<u>Project Reporting</u>

➲ You should be able to download or receive project reports for all aspects of the project. The reports should reflect overall project summary, the tasks that are pending or completed, the contact information for each team member, and the discussion threads or project updates.

BENEFITS OF UTILIZING THE RIGHT TOOLS

The benefits of utilizing the proper tool during a Web-related project are immense. In fact, if you do not use this type of tool you may be handicapping the project right from the beginning. Here are a few of the benefits of utilizing a project management tool:

➲ Enhanced communication between team members

➲ Hitting milestones successfully and on time

➲ Better project team management

➲ Cost effectiveness

➲ Quality completed project

Utilizing the right project management tools improves operational competencies, leading to real ROI. Companies are continually searching for ways to do more with less. All project tools should be viewed strictly by how they improve the final product. The correct tools will save every member of your staff time, and enable your project to meet the expectations set out in the beginning.

3.4 DEVELOPMENT PROCESS

Purpose: **Understand the proper process for developing Web projects.**

Key Points: **Evaluate the proper development process and implement the steps where necessary.**

Every new Web project poses challenges, but having a well-defined development process can help. Lucky for you we've made all the mistakes via trial and error. Our development process was created and refined from time in the trenches, completing complex projects that entailed sophisticated problems. In section 3.2 we covered the project management process. In this section we get into specifics of the development process. If you complete Web projects in–house, make sure you incorporate some of the following ideas into your plan. And if you use an outsourced vendor, make sure they have an existing process.

OUR PROCESS

Developing a large Web project is a process that may affect your business in several ways, including budgetary choices, personnel utilized, and sales results. One problem is that too many new Web projects begin as off-the-cuff, disorganized messes. The projects are often started by small groups within organizations, working in isolation to accomplish their own agendas. The results can be devastating.

The process should include a step-by-step approach, anticipating problems. The process should be clear and simple so that all involved understand how the actual pages will be completed.

Figure 3.4: Typical Development Process

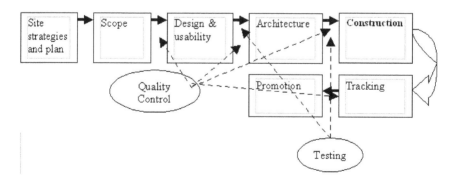

As you consider the development process outlined above, note that the construction of the pages that make up the Web site is one of the last things that takes place in a well-designed project. Consider each step in the process and its impact on your specification plan. Think before you act, and make sure you have the organizational backing, budget, and personnel resources you'll need to make the project a success.

POTENTIAL BOTTLENECK AREAS

As you review the above diagram, you'll recognize problem areas (if you worked on a web project) and we'll point out the areas where the process slows down (or can become stalled for good). The bottlenecks tend to occur when feedback is required from the client or when content is delivered late. Acerbating problems can occur in the testing stage between architecture and construction. If a development team transitions inappropriately during this stage, then the project timeline will be displaced. Make sure your internal staff and the development team discuss potential bottlenecks and make a plan to get around them.

3.5 COMMUNICATION PROCESS

Purpose: **Understand the process for communicating during a project and overcoming expected obstacles.**

Key Points: **In this section you will learn what can hamper the communication process during a Web project. You will understand what to expect and how to overcome the obstacles so that you end up with an open and productive communication process.**

Communication has a big impact on projects, from start to finish. The human art of communication becomes amazingly difficult during Web projects. There are many reasons why it is difficult. Common mistakes can be overcome easily and simply when each team follows a well-designed communication process.

There are many factors that we covered in the project management process that will help, however, what can be used during the project to make the process better? Furthermore, how can a process help make the completion of your new Web project successful?

THE PROCESS

Web projects involve many parts working together properly. Can you imagine trying to complete all the different parts of the project with no communication?

You must utilize a process that involves initial ground rules. At the very first meeting, discuss how the communication will take place and by whom. Identify how often, and what will be done if the communication breaks down. Ensure that each team member understands each other's ideas, issues, and challenges. As the project moves forward, different problems will arise.

Figure 3.5: Communication Process

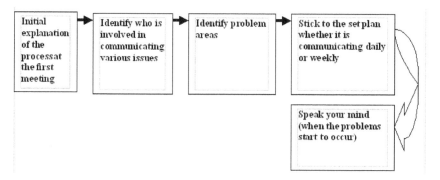

No process is wrong, but the key is to have the process in place. Describe it to everyone involved in the project. This way you get away from the assumptions (we all know what assumptions do).

3.6 BUILD PAGES

Purpose: **Understand when to build out the pages.**

Key Points: **In this section, we cover when to build out the pages.**

It probably seems strange to talk about building Web pages in the third phase of a development project. Besides, isn't building the Web pages the main point of any Web project? Building the pages is the easiest part of any Web project. The standard technology is the same whether you use HTML, DHTML, Flash, XML, etc. - the code is simply code. What matters are the personnel implementing the code and their ability to insert the code without error.

THE TIME IS RIGHT

Now that you've completed roughly 36 parts to your Web plan, you're ready to build the pages. In fact, if you've done the previous phases correctly, this part of your project is the most straight-forward. There is nothing left to do but have your team build out the pages. However, make sure you monitor the process with established quality control practices and that the personnel building the pages have the ability to finish the Web project without errors within budget and time constraints.

PROPER CODE

Programmers often get caught in the "any code will do" category, meaning quality is overlooked for speed. Proper code for a web project is the same as proper concrete for the foundation of your house. If it is weak and improperly implemented, you'll be surprised at the amount of extra time and money you and your team will spend trying to solve the myriad of problems generated by quick coding fixes.

The following criteria will help your team understand and judge the quality of your web project's code:

Use less code
➲ Lavish code that incorporates slavery to maintain it isn't fair for anyone involved and increases the cost factor.

Reduce unused functions and diagnostic statements
➲ They can complicate the project and increase development time

Code must be maintainable
➲ You don't want to reinvent the wheel with each add on or improvement to the code

Using existing libraries
➲ Starting from scratch is often necessary but not always; be absolutely sure the code you reuse is of good quality

Write in a standard language
➲ Something standard and universally accepted helps

Write test code
➲ Test before you log tons of hours with bad code

Improve performance
➲ The rule of thumb is that 1% of the code takes up a certain % of actual runtime. With larger, more complex applications this run time is increasing.

Documentation

➲ Document your business logic and structure with the code so that the next person can see what you were trying to accomplish and how you came to your conclusions.

3.7 QUALITY CONTROL & TESTING

Purpose: **Understand the process for quality control.**

Key Points: **In this section you will learn a process for controlling the quality of your end product. Evaluate different options for testing procedures.**

When applied to Web projects, the term "quality control" describes working to improve the quality control process. Indeed, you must understand and enforce the quality control standards and processes used to produce the end product. This includes its technical components, architecture, and content. If you have the right quality control process in place, a new Web site should see a decrease in bugs (defective programming) and an improvement in site usability and performance.

HOW TO IMPLEMENT

Quality control should enhance the user's experience, helping the developers to design a site or application with the users in mind. Brainstorm on all the factors that can go wrong – what will we need to overcome, other than the traditional problems?

Focus on understanding and tracking the problems. Proper quality control for a Web project involves an interaction with the internal development team and outsourced vendor, at several levels, in order to find bugs and provide clarification. The process should catch and report problems and direct the issues to the right person for resolution. The test is simple. The code passes or fails, and is then referred to the team to identify who is responsible and figure out how it will be fixed. Test again and then solve until completion.

A properly established quality control process contains a reporting system, typically Web based, and trackable. The system should enable the development team to recognize problems, report them via the Web, and track their resolution. Make sure you have a clear understanding of how the problems will be dealt with, who will be responsible for what, and the expectations regarding the timeline on getting the problem fixed.

QUALITY CONTROL CHECK LIST

Before the final site or application is launched live, make sure you test for the following:
➲ Compatibility with the hosting platform
➲ Database issues
➲ Bugs in the code
➲ Download speeds from 56K and up
➲ Spelling and grammar
➲ User Interface usability
➲ FTP access
➲ Administration areas (do usernames and passwords work?)
➲ Security issues

Launching a new Web project without completing the proper quality control and testing procedures can cause embarrassment and in some situations, the loss of employment. In the end, make sure all deliverables have been accomplished and everything works properly before the new site or application goes live.

DEBUGGING

There are many new methods and software applications to help debug the code in your project. These techniques have improved over the years in response to changes in user profiles, programming languages, and implementation techniques. Make sure the development team you work with describes their procedure and systems for the debugging process. Bugs in code occur and are very common. In fact, you'll rarely if ever see a web project brought to the testing phase without bugs. The key is not if bugs occur, but how they'll be dealt with once they are recognized.

COMPLETION OF MODULE 3

You have now completed the Construction phase of your Web plan. Refer to Module 5 for the outline and details of a properly constructed Web plan. The construction phase is simple if Phases One and Two are completed properly. Make sure that the personnel, whether internal or external, have the capabilities and drive to finish the project without errors and on time. After completing Modules 1, 2, and 3, your new Web project is complete, meaning the actual pages are built and the site or application is live. At this point, you have a fully functioning web site or web application. Make sure you continuously ask for and receive feedback from the users. Next, we'll help you match the online marketing tools to the objectives you established in Module 1.

MODULE 4

DRIVE

4.0 Drive

The fourth phase "D" in the IDS^{SM}, or the A, B, C, Ds of development is "Drive." This is a reference to not only *driving* traffic to a newly created, newly rebranded, or newly modified Web site, but it also refers to *driving* your business forward. The online world has evolved so much that marketing via the Internet today is about much more than building a Web site and driving people to it. It's about anticipating your target audiences' online behavior, showing-up where they may be within the world-wide-web (i.e. showing up in the results of a search query, etc.), getting them to turn their head your way (i.e. getting them to click on your listing in the search query) and serving them relevant content that either:

a) gives them pertinent information relative to what they may be seeking (i.e. a press release or blogging article served to them as a component of relevant keyword terms); or

b) Encourages and/or incentivizes them to allow you to develop a relationship or rapport with them via—in many instances these days— a two-way dialogue.

According to a study by Ernst & Young, it is estimated that 55% of consumers who have purchased furniture or major appliances researched online before eventually purchasing offline. In 1999, only 15% did online research for this type of purchase. Businesses spend significant time, money, and resources grabbing and then keeping the attention of website visitors.

The reality is that most firms are losing their visitors the moment the visitor hits the website. In fact, Forrestor Research has stated that 70% of online customers will drop brand loyalty and go with a competitor if they do not receive an appropriate or timely response once on the web site. As you go through Module 4.0, you will quickly understand how to match Web-based promotional tools to your overall business objectives. This is critical to the success of your project. The ideas and tools for promotion are a large part of your overall Web plan.

Thus, to *drive* could entail any combination of the following online media/marketing tools or social mediums, also commonly referred to as digital media: search engine optimization (SEO), paid advertising (pay-per-click), email marketing, blogging, podcasting, online video, message boards, chat rooms, online and social networks.

To develop your plan to *drive*, you must start with the end in mind. In Phase "A", we covered how some of the promotional tools can help make business results obtainable (i.e. an ROI). Part of establishing a potential web and/or online ROI is building a solid online marketing & sales strategy, which includes an online marketing plan filled with online media tactics like those mentioned earlier. You need to understand how all your marketing tools will work together, both traditional media (i.e. direct mail, print, radio, television, etc.) and online media. There are plenty of books on traditional marketing tools so we will not get in-depth here. However, when it comes to the mix of online marketing tools, the key is to match the tools to fit your strategy, target audience(s) and your objectives.

Most books and Web firms cover a promotional phase strictly from the tactical view point (search engines, email, and traditional marketing). Many focus on just driving traffic. We tend to ask "what type of traffic do you want?" It is definitely quality over quantity these days. As we talk about the *drive* component, keep in mind that this phase involves attracting your best potential customers, keeping existing clients happy with timely communication, and corresponding appropriately with partners and vendors.

This module will assist you in laying the groundwork for a detailed, online marketing & sales strategy that will reach your target audience(s) with the right message and allow you to achieve your goals and ROI.

4.1 ONLINE MARKETING & SALES STRATEGIES

Purpose: **Understand the elements of an online marketing & sales strategy and how to develop one; and how to determine and coordinate which online marketing tools to include as part of your implementation plan.**

Key Points: **In this section you will learn a process for evaluating and implementing the correct online marketing and sales strategies. You will understand how to establish the strategies and complete the tactics, matching the options to the overall marketing objectives. You will feel confident in implementing specific tools because you will have defined the strategies, understood the limitations, and qualified the budget and resource requirements.**

Your online marketing & sales strategy should be a component of your overall web strategy and tie into your grander marketing strategy—depending on how your organization approaches your business from a strategic perspective. That said, the online space is very technical in nature and subsequently has often been assigned as the responsibility of the IT department. We recommend that management own the company's web strategy and the IT and Marketing departments equally be assigned to provide resources to help execute the strategy. As mentioned in the opening of this module, begin developing your online marketing & sales strategy with the end in mind. Doing so will have a significant impact on the structure of your Web site: what it does, who interacts with it and how they behave while there.

The key to a proper online Web strategy is a balance of all the available tools, and matching each tool to an ROI indicator. If online marketing brings the users to the portal/web site, then online sales becomes your closing tools to turn prospects into paying customers.

DEFINE YOUR STRATEGY

Begin by determining what you want the web site to do for your business—both now and in the future (2 years from now, 5 years from now). What do you want to accomplish? This should be easy as you should have addressed this same question in Phase "A" earlier before you set out to build your web site. Do not lock yourself into any legacy thinking here—be open to grandiose ideas—ideas that today may sound like they may never happen. In brainstorming this internally, follow the philosophy that there is no such thing as a bad idea. Considering how fast technology is evolving in the online arena, you'll be surprised to watch some of the ideas that are not feasible now become possibilities or realities. This exercise should elicit an online marketing & sales strategy statement—which essentially describes your definition of success relative to your online efforts. Below are a couple examples of what a strategy statement might sound like. Note how, generally speaking, they each state an overarching goal and explain broadly how that goal will be achieved.

➲ To earn a reputation as one of the top service providers through extensive communication of our accomplishments, service-levels and customer satisfaction ratings among our targeted prospects and through the continued provision of exceptional customer service.

➲ To be the "go to" online source for information relative to our products and industry by establishing ourselves as experts through the publishing of papers, articles and opinions, etc; and promoting such materials to our target constituents to drive online traffic to our Web site and subsequently increased online sales.

SET OBJECTIVES

Once you know what you want to accomplish online and you've defined what success looks like, you are now ready to chart a course to get there. Unlike the line made famous from the movie *Field of Dreams* in the 1989 film where Ray Kinsella (Kevin Costner) heard voices telling him, "If you build it, they will come", which will not work on the web. Today there is so much content online that getting people to find yours and—when they do—give it any attention is your biggest challenge. You put the stake in the ground with your strategy statement and articulated what you wanted to do; now you need to create your implementation plan and set some objectives to measure your progress along the way—this way you'll know when you get there.

Your marketing & sales strategy will be the equivalent of a GPS system in a car. It will enable you to get to your targeted destination. With a GPS system, if you take a wrong turn or if there are road closures, in theory the system will re-route you to your destination via an alternate route. The same is true with your marketing & sales strategy. As opportunities present themselves, or your competitive environment changes, or your overall business priorities change, your strategy can evolve and your course to success can be modified. Your marketing & sales strategy will be a great road map from which to make business decisions.

In setting your objectives, they should tie into your strategy so that by meeting one or all of them you will come closer to the full execution of your strategy and the achievement of your overarching goal. Be as specific as possible and ideally, make the objectives measurable—having a metric attached makes it very black-and-white as to whether you achieved it or not. Also, make sure that your objectives are realistic and obtainable. Setting an unrealistic bar will set you up for failure. Below are examples of what a few of the corresponding online marketing & sales objectives to the aforementioned strategies might sound like:

- ➲ Obtain a customer satisfaction score of 95% or higher from each customer, based on our annual customer satisfaction survey results each year.
- ➲ Expose each of our targeted prospects to our list of accomplishments and exceptional service-level history at least 3-7 times throughout the year (via various online marketing channels).
- ➲ Experience a 10% increase in requests for proposals this year from our targeted prospect list.
- ➲ Generate 25% more online sales among our targeted prospects by 3Q.
- ➲ Increase online traffic to our Web site/store by 25% as measured by natural web traffic (i.e. not including pay-per-click).

INDENTIFY ONLINE MARKETING TACTICS

With your strategy statement and corresponding objectives in place, you are now ready to start physically mapping out your implementation plan. To do this, you need to begin by recalling the user profiles that you created in Phase "A" with regard to your targeted audience(s) and visitors to your Web site. Phase "A" focused on who they were and how they will behave while on your Web site. Now you need to put yourself even further in their shoes and try to understand their overall online behavior. Ask yourself questions like:

- ➲ How do they go about finding information online when they need it?
- ➲ What do they read online?
- ➲ Do they only surf the web, or do they like to subscribe to RSS feeds, blogs, podcasts and the like?
- ➲ Are they willing to register to receive emails from trusted sources if they promise them valuable and insightful information?
- ➲ When they aren't thinking about your product or service, but are online, where are they? What Web site(s)?
- ➲ How does their online behavior compare with what you desire that they do online on your Web site (i.e. your call-to-action)?

This list is certainly not exhaustive, but gives you an idea of the exploration you need to do. The answers to questions like these will give you a basis for selecting the online marketing tactics to execute your strategy.

Once you understand your target audience(s)' anticipated behavior better, you are ready to begin to select the elements of your implementation plan.

Outside of your Web site—obviously—your plan will include applicable online marketing tactics such as the following:

➲ Ecommerce applications
➲ Search engine marketing (SEM)
➲ Email marketing tools
➲ Online advertisements
➲ Social media (Blogs, Podcasts, Social networks, etc.)
➲ Affiliate marketing/strategic alliances
➲ Online sales department/customer resources and tools

We address some of these tactical elements in more detail later in the module. However, the key to an effective implementation plan is the proper balance or mix of all the available tools, and matching each tool towards achieving your objectives—and ultimately your strategy. If you can't match a tactic as a contributor towards meeting an objective, then you know that tactic probably shouldn't be part of your plan.

AUTHOR YOUR MESSAGE

The next step in preparing your implementation plan is to make sure you know what you are going to say. Each online tactic that made your online marketing mix in your implementation plan will require a message. If you already have a traditional marketing plan in place (i.e. direct mail, print, newspaper, radio, television) and you've carefully crafted your message strategy for the corresponding tactics, then you will certainly want to maintain consistency with your online mediums and continue to leverage that message with your online marketing tactics. However, if you haven't gone through that exercise, while creative messaging is out of the scope of this book, here a few tips for developing your messaging strategy and overall messaging points:

➲ Define your product/service offerings
➲ Compare yourself to the competition
➲ Identify why you're different
➲ Craft your marketing message around your differentiating points

4.2 ECOMMERCE APPLICATIONS AND STRATEGIES

Purpose: **Understand how to effectively implement ecommerce tools on your site and properly promote the products.**

Key Points: **In this section you will learn how to implement an ecommerce strategy. We cover the potential tools,**

the shopping process, and the strategies to drive traffic. Any good ecommerce strategy will increase sales and margins.

While ecommerce applications can be a component of an online marketing & sales strategy and implementation plan, when appropriate, to develop an ecommerce application requires a separate ecommerce strategy. Thus, the ecommerce strategy drives the ecommerce application that is ultimately a component of the larger online market & sales strategy that feeds into the overall web strategy, which then ultimately ties into the fundamental business strategy (Recall the *Strategy Hierarchy* below).

Ecommerce has transformed the Web over the last 10 years from a brochure tool to a viable way to reach revenue and profit goals. According to Webopedia, the definition of ecommerce or, *electronic commerce,* is business that is conducted over the Internet, using any of the applications that rely on the Internet, such as email, instant messaging, shopping carts, Web services, UDDI, FTP, and EDI, and others. Electronic commerce can occur between two businesses (B2B) transmitting funds, goods, services and/or data, or between a business and a customer (B2C). For the purposes of this module when we refer to ecommerce, we're referring to it strictly as it relates to the selling of goods, where a financial transaction takes place in real time over the Internet. In order to implement the right strategy and application, you must understand the market opportunity, the tools, merchant accounts, and promotional methods. In 2007, Forrester Research estimates $157.4 billion in retail transactions were made online and they predict that it will grow to $271.6 billion in the next four years representing an average annual growth rate of 17.5%. While this projected annual growth rate is down from the 24.7% that was averaged between 2003 and 2005, we note that it's still almost a 20% growth rate, which is significant in any industry.

In fact, today ecommerce is internal, external, can come from blogs, via linking social networks, video connections, and about every other imaginable digital method. For our purposes, we're tackling ecommerce strictly for the selling of goods, where a financial transaction takes place in real time on the Web. In order to implement the right tools and strategy, you must understand the market opportunity, the tools, merchant accounts, promotional methods and social media connections.

Figure 4.3: Strategy Hierarchy

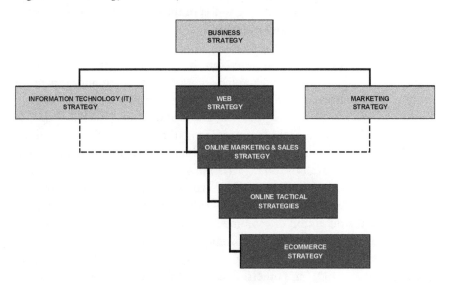

DEFINING AN ECOMMERCE STRATEGY

The first step in implementing an ecommerce program is to identify which products you will sell. For our purposes, we're covering products that can be bought online and then either downloaded or shipped to the customer. You don't have to sell everything – some products lose money when sold on the Web while others have impressive margins. Next, you'll need to review the shopping process options. This is usually referred to as the shopping cart. How do you pick the right option?

IMPLEMENTING AN ONLINE SHOPPING PROCESS

Choosing an online shopping cart that fits your needs is a critical decision. Similar to a brick and mortar retail establishment, the shopping process on your Web site is your only opportunity to impress potential buyers. If the shopping process is hectic and difficult, a retail store can go out of business, very quickly. This is true for your Web site's shopping cart, as well. If the user has problems, he's gone. In fact, the most important part of the shopping cart is the customer's private information

and credit card numbers. If you do not deal with this information in the right method, you'll lose sales. *Security is paramount.*

There are several options when it comes to shopping carts. They range from Web-based systems where your site simply links to the cart, or on-site systems that require programming to integrate the cart into the site. The three main categories include merchant services like eBay, hosted shopping carts, and integrated shopping carts.

Merchant shopping carts like eBay require you to utilize your own merchant account and gateway. A merchant account is the bank account that enables your site to take credit cards, while the gateway enables the transaction to be completed over a secure Internet connection. A site like eBay will allow you to use their system (merchant account and gateway), but you pay higher fees. A merchant service is perfect for companies that are just starting their business or do not have a significant amount of transactions.

Hosted shopping carts are similar to a merchant system except they include the merchant account, gateway, and shopping cart. They typically charge an upfront set-up fee and then a small fee for each transaction. This is also good for a small business just starting out. The negatives of this choice include a templated cart (your image may not match their offerings) and few functional options other than the base plan.

An integrated shopping cart is a full custom, programmed cart. You receive all the "bells and whistles" and can often leverage many different technologies to up-sell or cross-sell various items. This process is the best option; however, it will be the most costly.

Once you've chosen the best option for your ecommerce process, you will need to run tests with usability (How easy is the cart to use?), marketing tools (How do you get the user to the cart?), and up-sell or cross-sell opportunities (How do you set up accounts and when do you offer more options?)

COMPLETING THE SALE

Now that you've chosen the type of shopping technology you'll use, it's time to review the keys to increasing your online sales and making sure users complete the transaction. Here are some helpful tips:

Make the shopping experience simple:

➲ Make sure the products are easy to locate, that pictures and graphics match the product, and that product descriptions are easy to read.

Provide detailed explanations:

➲ Explain both your shopping process and other policies in detail. Show a diagram and explain in words, how the person can, and will, shop your site. The more specific your directions are, the more shoppers you will retain.

Common sense check out:

➲ Allow the customer to purchase products and/or fill up their shopping cart and then ask for payment. During the payment process, ask for detailed information about the customer. Don't ask for this information before they enter their credit card number and don't over do the questions on the detailed information. Their dog's favorite toy might not be appropriate – stick to the basics – contact information and customer feedback.

Multiple shipping options:

➲ Use a shipping company that has on-line order tracking. Make it easy for your customers to use this.

Dividing sections:

➲ Make sure your products are divided into categories that make sense. For example, you will want all the wood products to be together, and separate from the metal products, or your consulting services to be separate from your financial services.

Accept credit cards:

➲ The great advantage of on-line shopping is its immediacy. The entire transaction should be completed in just a few minutes. Customers will simply leave and go elsewhere where the need for instant completion of the transaction can be fulfilled if your process takes too long, or in the extra time they have before completing the transaction, customers may change their minds about purchasing from you. Also, remember that credit cards are by far the easiest way for non-U.S. customers to order directly from the U.S.

Confirm Orders:

➲ Make sure the customer not only receives a page saying, "Your order has been received" (or something similar), but also the ability to print out a receipt for their records and receive a confirmation email.

Provide Help Areas:

➲ Create a help area similar to FAQ (frequently asked questions). This area should include the FAQs along with methods to get in touch with you, and a link to your live chat tool, if you have one. Address problems that may come up and solve them online whenever possible.

List Important Information:

➲ Your customers should be able to place an order without a hassle. Make sure they don't have to fill out three pages of forms before the order is accepted. Receive important information for your database, like name, address, preferences, zip code, etc. *after* they have placed their order. The fewer steps and shorter the process, the more likely the customer is to complete the sale.

Remember International Customers:
- ➲ Don't lose orders by narrowing your market just to your home country. Make sure your order form takes international zip codes and provides a valid payment processes. Many sites reject orders if the zip code is not a valid U.S. zip code. International customers have money just like Americans. Many will find your site on a search engine and if your interface is "internationally friendly" they may order. At least give them the option.

Implement one click ordering:
- ➲ Amazon trademarked this process, but you can use elements like storing the customer's data (be careful regarding the credit card information – Disclaimer: there are ways to store customer data without displaying the credit card number – please see a professional consultant regarding this process). The user confirms the data, shipping information, type of payment and then clicks to order. The rest is automatically completed.

Customer should confirm their order before buying:
- ➲ The last step in the process should reflect the customer's order, summarizing the items with a total. The shipping address should be confirmed and allow them to either place the order or cancel. Make sure the order number is correct and that the customer knows how to track the order.

Send a confirmation e-mail:
- ➲ According to Jacob Nielson, e-mail confirmations should be brief, tell the user what they bought and their order number. Also, the form should discuss how to clear up a mistake, buy more product, and learn more about your firm (in a brief manner).

IMPROVING THE PROCESS

The key to a well-defined ecommerce strategy is understanding what can go right and wrong in the process. Here are some ideas for improving the process:

Anticipate problems:
- ➲ Your first step in implementing online transactions is to brainstorm on all the challenges. Every online transaction process has problems. The key is to understand each one and either find a solution or create a process for handling the problem. For example, what happens if the credit card number is entered with 17 digits? Does the site crash? Or, is there an explanation page that pops up to direct the user back to the problem field?

Use e-mail addresses to identify shoppers:
- ➲ Asking for the user's email address is a simple way to create usernames and accounts. The user's email address converts to a unique username that the customer can remember and the format is standard with a name@emailaddress.com.

Highlight the required fields:
- ➲ Order forms need to have a symbol next to required fields and not allow the user to go to the next step in the shopping process unless they input information. If you don't point out your required fields, the user becomes frustrated when the form is not filled out correctly and they have to go back to the page in order to fill out the correct fields.

Shopping cart vs. order form:
- ➲ Many Web site owners and managers are unsure of when to use a shopping cart tool or build a catalog, and when to use a simple order form. There are two keys to this decision: the number of products in the offering, and the number of products purchased by each customer during the purchase process. For example, analyze whether your customer will buy just a hat or purchase a pair of socks with shoes and shoe polish. Typically, if you have less than 20 products, an order form works well. If your customer orders more than two items at one time, a shopping cart mechanism will work best.

Dynamic shopping cart:
- ➲ If you have a lot of products, or make changes frequently, then you will need to use dynamically generated pages. However, many search engines may not index. Design issues come into play with this type of cart in order to offer a seamless appearance to your site.

Static shopping cart:
- ➲ A small amount of code is added to the pages containing your products. This type of cart is great for optimizing each page for the search engines. You will need a maintenance agreement with your development firm when you choose a static cart, and a new page must be created each time you add a new product.

Understand the payment process:
- ➲ Merchant accounts range from no set-up / no monthly fee (but high per-transaction charges) to high set-up / low monthly fee (but relatively low per-transaction charges). The kind of account you choose should depend on how many orders you expect and what of professionalism your visitors expect. Be careful of merchant services that have your visitors leave your Web site to complete their transaction. This customer data is often sold. If you expect a lot of transactions, you'll probably want a merchant account with a bank, offering the bank's

own payment process/tool. This assures merchant account/payment processes will be compatible. Don't be conned into signing up for a low-cost Web-store that commits you to high cost or restrictive payment processing.

Commitment to Privacy:

➲ All the transactions on your site should be protected and secure. All your customers' personal information should be protected and never sold, unless prior permission is given by the customer.

Site Performance:

➲ You don't want a sluggish site. Before signing up, check out some reference sites to see how they perform. Remember to test out your Web site and ordering process at different connection speeds, different screen resolutions, and with all the commonly used Web browsers.

Resolve problems immediately:

➲ Your customers must have access to fair, timely, and affordable methods. To resolve transaction problems, clear and sufficient information should be provided so consumers can make an informed choice. Vendors should take reasonable steps to ensure the consumer's choice is informed and intentional.

Outsource transactions:

➲ Unless you have very particular server-side requirements, consider very carefully before deciding to use your own server, rather than a hosted solution. If you have many hundreds or thousands of items to sell, you should really be looking at solutions which link in to a back-end database (or at least have an import facility).

Timing:

➲ Get the order to the customer ASAP. The Web is about <u>seconds</u>. Not days. Larger competitors will offer free shipping, often with same day delivery, while smaller competitors will offer more unique choices. Make sure your delivery system and the timing of getting the product/ service to the customer is at the same level of quality as the product and customer service.

STRATEGIES TO INCREASE ONLINE REVENUE

There are specific ways to build sales momentum, making your online ecommerce explode. Most tools and ideas do not take extra budget, but rather time to evaluate the relevance of implementation. The right technology with the right process does one key thing – improve the user's experience. The following tips can help improve and deliver e-commerce results:

➲ Improve the look and feel – create trust

➲ Make a personal connection – the user wants to be cared for

➲ Make honest offerings – tell the entire story so that the user doesn't have to ask for it

➲ Find more strategic partners – increased sales force

➲ Cross promote – most sites miss this step

➲ Continuously push the user to the check out page

➲ Establish delivery policies that work for your system – don't over promise and under deliver

The ecommerce strategies that you decide to move forward with will involve many parts of your business, both from a technical standpoint and a process standpoint. The information overload can cause headaches. However, you can succeed if you divide the process into three to four sections and match your overall business strategies and subsequent online marketing & sales strategy to the specific ecommerce tool(s).

SIDE BAR: 10 ECOMMERCE EVENTS THAT SHOOK THE LAST 10 YEARS

E-Commerce has been an extremely fast growing industry and within the next four to five years is projected to represent almost 10% of all retail sales. Below is a list of the top 10 e-commerce events that shook the last 10 years—compiled by the Software and Information Industry Association (SIIA)—based on voting by industry experts and policymakers (July 2007).

1. Google (Sept. 1998)
2. Broadband Penetration of US Internet Users Reaches 50 percent (June 2004)
3 .eBay Auctions (Launched Sept. 1997)
4. Amazon.com (IPO May 1997)
5. Google Ad Words (2000)
6. Open Standards (HTML 4.0 released - 1997)
7. WiFi (802.11 launched - 1997)
8. User-Generated Content (YouTube 2005)
9. iTunes (2001)
10. BlackBerry (1999)

4.3 SEARCH ENGINE MARKETING

Purpose: **Understand the benefits of search marketing and how to determine what search marketing services**

you should include in your implementation plan (if any).

Key Points: **In this section you will learn a process for evaluating and implementing the correct Web-based promotional tools. Just because it exists does not mean you should use it. You will understand how to set budgets, evaluate options, and make adjustments.**

Search engine marketing (SEM) refers to the overall effort of positioning your web site to help ensure that you are listed within search engines when a user conducts a query on a search engine by a keyword or phrase that is relevant to your business or industry. It is the art of enabling people to find your Web site on the web. Note that this is an art and not a science. SEM includes search engine optimization (SEO) and search engine advertising (pay-per-click). Most people learned of the tremendous power of SEM via the Google initial public offering (IPO). Google has taken the SEO and pay-per-click model to new heights—so much so that it has effectively become a verb. "To Google" something is to conduct a search on Google.com for a particular word or phrase. Unless you are President Bush who will forever be remembered for referring to Google as a noun. In a October 2006 interview on CNBC's show *Closing Bell* with Maria Bartiromo, he responded to a question on whether he uses Google by indicating that, "*One of the things I've used on the Google is to pull up maps*" to see his ranch in Texas. There are certainly other search engines besides Google, but as you will note in the ranking graph below, Google accounts for nearly 60% of the online searches.

According to Zoomerang, nearly nine out of 10 online shoppers said they found retail sites through search engines. About the same number of respondents said they were inclined to visit sites they had never heard of because of a high search ranking. This is good news if you sell products directly from your web site. However, what is the situation if you sell a service from your web site? The latest findings show that over 50% of the web users research a firm via the web before making contact and/or a purchase. How do they find those firms? They find them by using a search engine.

In 2006, the PEW Internet and American Life Project reported that out of the 94 million American adults that search online in a typical day, nearly 60 million people (64%) used a search engine. According to Marketing Sherpa, SEM was $8.7 billion industry in 2006, which was up 26% from the $6.9 billion that was spent in 2005. While growth projections vary, some estimates like one from Piper Jaffray® in 2005, projected a 37% CAGR per year for five years, which would make the industry a $33 billion industry by 2010. Thus, the bottom line is that the SEM market is going to continue to grow and it's likely that if you aren't promoting traffic to your Web site, your competition is promoting it to theirs. In this section, we'll cover the different elements of SEM so you can better ascertain which components are right for your online marketing & sales strategy and plan.

Figure: 4.4: Search Engine Rankings – Five Major U.S. Search Engines

According to a study conducted by comScore, as of September 2007, below is the U.S. market share based on the 9.4 billion searches that are performed in aggregate at the five major search engines.

Search Engine	September 2007
Google	57.0%
Yahoo!	23.7%
Microsoft (MSN)	10.3%
Ask	4.7%
Time Warner	4.3%
Total	100.0%

Notes:
1. Based on 9.4 billion searches at home, work, and universities.
2. Data are based on the five major search engines, including partner searches and cross-channel searches. Searches for mapping, local directory, and user-generated video sites that are not on the core domain of the five search engines are not included in the core search numbers.

Source: comScore, 2007

When embarking upon SEM, essentially you will need to begin by assessing the traffic that you are currently getting to your Web site and determine how quickly you need that number to increase (see section 4.9 *Track & Adjust*). Naturally, we all want more traffic to our Web site(s), but with SEM that traffic comes at a price, whether it's a price you pay in doing it yourself (i.e. time, learning curve, etc.), or the price you pay to hire a search professional to help you build traffic. Regardless, you will want to consider the following two primary options as it relates to SEM:

1) <u>Search Engine Advertising (Pay-per-click)</u>
   This option is the shorter-term option for more immediate results. You can initiate a pay-per-click program and almost immediately see your Web site come up within listings.
2) <u>Search Engine Optimization (SEO)</u>
   This option is the longer-term initiative as it takes longer to see results. Within SEO, there is also press release optimization that is quickly becoming a must for companies online, and video search optimization, which has begun to rear its head and will only continue to increase as online video continues to grow.

We speak about each option more below, but generally—especially with new Web sites—you should engage in a combination of the two. Initiate both options at the same time and while the SEO is seeding itself, traffic is being driven to the Web site via pay-per-click results. As SEO results begin to appear, the pay-per-click program can be scaled back accordingly.

While each option is featured more below, it should be noted that entire book series have been authored on the art of SEM. The information below is simply designed to give you an understanding of the importance of the tactic, guide you through determining what SEM programs are right for you and provide you with some cursory knowledge and tips for implementing an SEM program(s). Additionally, we should note that search engine technology evolves so quickly that some of the information contained within this section could very well be obsolete within a year, which is actually a benefit to hiring a search professional because it's their job to keep up on the changes, updates and advances in technology.

KEYWORDS

A fundamental component of either pay-per-click and/or an SEO program is keywords, or even keyword phrases. Either program requires the selection of the right keywords or phrases that will appeal to and achieve the targeted visitors that you may seek. You can determine these based on your knowledge of and expertise in your industry. You know what terms your prospects and prospective visitors are using when conducting queries via search engines within your industry. If you don't, then you should take some time to inquire and understand what people are using to search. You can then utilize software, or consult with a search professional to run an analysis on your selected keywords or phrases to validate their potential. Analysis reports can indicate how many times your selected terms are entered as search terms. Obviously, if the report comes back and your initially selected keywords have low query numbers, you might want to rethink using them for pay-per-click or optimizing them since they may not garner you any traffic.

A good web developer will advise you on the importance of giving thought to and determining your keywords and phrases when you initially pull your Web site content together—i.e. the Blue Print phase within this book (phase "B").

SEARCH ENGINE ADVERTISING

Search engine advertising is also known as pay-per-click advertising. Pay-per-click search engines, such as Google and Yahoo!, are search engine Web sites that list results in a certain area of the screen (separate from the organic listings—usually the first couple of listings at the top of the page—called the "one box" and then the listings down the right hand margin of the screen) based on the advertisers that were the top bidders for that keyword or phrase. It requires setting up an account online with the desired search engines so that you can manage your pay-per-click program through a password protected back-end portal. Google and Yahoo! are the two primary pay-per-click search engines because Google's results feed MSN's search engine and Yahoo!'s results feed AOL's search engine with results.

HOW IT WORKS

The pay-per-click advertiser selects the keywords or phrase, sets bid maximums, daily budgets, geographic and/or day-part parameters—and then they let the users click away. The advertiser that bids the most for a keyword or phrase gets its listing to appear first in the results list; the second-highest appears second, and so on. If a user clicks on a paid advertiser's offering to go to their Web site, the pay-per-click search engine charges the advertiser.

Typically, Web site owners utilize a pay-per-click provider, either directly or through a search professional who manages the pay-per-click process. The prices for keywords and the click-through rates vary according to how competitive and/or popular your keywords might be for your industry. For example, the price-per-click for the word "computers" might be $10.00 per click, while the rate for the words "Asian tap dancing" might be $0.05. Obviously, the more popular a keyword term is as a potential search term by your target audience, the higher it will likely cost for a click. Subsequently, as your competition realizes this too, then as you each bid on the term, it drives the per click price even higher.

One of the newest tools is the Google AdWord tool. When utilizing the AdWord tool, you will typically be able to do a keyword match, which shows your ad when a search includes your selected words and phrase matching, or includes your selected phrases. In addition, an exact query match shows your ad when a search contains your exact keywords and no others, while a negative keyword match will not show your ad if a search contains words you select.

IMPLEMENTING PAY-PER-CLICK

The advantages of pay-per-click are quick results, controlled spending, and competitive leverage. Pay-per-click can be a great ongoing tactic for your online marketing & sales strategy and plan, or as we mentioned in the opening, it can be a great interim tactic while your SEO program is ramping up. Here are some tips to consider when embarking upon a pay-per-click program:

- ➲ Determine your keywords – for what words do you want to appear in search listings?
- ➲ Define your budget – determine what you are willing to spend on a monthly basis
- ➲ Determine your search engine(s) – from how many search engines do you want to buy clicks? There are literally hundreds—but your budget may drive this.
- ➲ Determine your dayparts – what time of day do you want to appear in search listings?
- ➲ Determine your geography – what part of the country/world do you want to appear in search listings?
- ➲ Determine your landing page strategy – where are clicks from your ad going to go? Your website or a special landing page with a special offer?
- ➲ Deploy the program

Note: While these are great directional tips to get you started, keep in mind that the pay-per-click industry is ever-evolving and can be fairly complex. While there are many advertisers out there who do manage their own pay-per-click programs, realize too that there are also many search professionals who are dedicated full-time to helping clients implement, execute and manage pay-per-click programs. Depending on your situation, the available internal resources you may have, your budget, etc., you will want to weigh whether it makes sense for you to go it alone, or enlist the help of an outside vendor.

SEARCH ENGINE OPTIMIZATION

Search engine optimization (SEO) is the process of designing pages, creating web content or text, and programming web pages for your Web site in a manner that will increase the probability that your Web site will appear at the top of major search engines when a visitor types in selected keywords or keyword phrases—also referred to as organic, natural or non-sponsored listings. To optimize a webpage is to manufacture the page for the best results in a directory or search engine, based on content, keywords/phrases, descriptions and links. Your Web site may look ecstatically appealing, but search engines don't care about design. How your Web site references targeted keyword terms is critical to how search engines will find you, particularly crawler-based search engines. In crawler-based engines, they crawl (or spider) through your Web site in an automated fashion, indexing it. The search engine then has a software program that, based on an algorithm, weighs different criteria about your Web site (such as keywords that it indexed, popularity, etc.) and displays listings per a user's search request. As mentioned above, SEO is a long-term approach as it can take anywhere from three to 18-months to see results within listings for specific and targeted keywords or phrases.

DEVELOPING CONTENT

Search engines look for content. **The** content or text on each page of your Web site is what most search engines look at, but each will have a different take. How do you meet the standards for all the different engines? You don't. You want to prioritize by visitors. Make sure the content is relevant to your targeted user and then let go. If you try to provide different content for each engine you'll lose your key visitors. Here are some brief guidelines:

Establish a clear purpose:
➲ Provide the purpose in the main page paragraph describing your
   company and products and services.
Stay on the subject:
➲ Make sure the content fits the subject and the keywords you've
   selected– stay on track.
Review for the word weight:
➲ Try not to repeat words in the content unless it has specific relevance.
Web visitors:
➲ Keep in mind who visits your website. What content will be important
   to each visitor and how can you help them complete your chosen "calls
   to action"?

Once you've determined the best keywords or phrases for your Web site and have incorporated them into the overall content of your website, you need to ensure that the Meta title, Meta description and Meta keyword tags are implemented on each page of your Web site:
Title tag

⊃ This tag is arguably one of the most important elements for generating strong search results for your Web site. It is the title and/or description that appear in search engine results pages (SERPs) that are clickable or linked. It typically includes your company name and sometimes a primary keyword or short description.

For instance, the megabrand Nike has the following title tag for its running home page: *nikerunning.com*

Description tag

⊃ This tag describes exactly what someone could expect to find if they viewed your Web site, or a webpage. It is the description or sentence that appears in each search engine results page (SERP) under the clickable link. This should contain keywords or phrases if possible, but each search engine gives the description tag different weight in their algorithms as it relates to ranking your Web site.

For instance, Nike has the following description tag for its running home page: *Use our innovative Training Log, Pace Calculator and other tools of the trade. Check out the latest NIKE Running products and services - everything you need.*

Keyword tag

⊃ This tag contains, in order of importance, the keywords people would more than likely search for when looking for either your Web site or information on your industry and the words that are listed in the content text of that particular page. You should have no more than 170 keywords and you should not repeat identical words/phrases. For example, you could have computer, computer keyboard, and computer system, but computer, computer, computer will be a red flag to search engines as spam. Note: even when using variations of the same word—such as in the example above—limit the repetitive word to only appearing in the tag 3-4 times.

For instance, Nike has the following keyword tags for its running home page: *RUNNING, NIKE, NIKE* RUNNING, SHOES, RUNNING SHOES, NIKE RUNNING SHOES, SHOX, NIKE SHOX, NIKE AIR, AIR MAX, PEGASUS, RUNNING GEAR, TRAINING, TRAINING LOG, TRAINING SCHEDULES, PACE CALCULATOR, SPLIT CALCULATOR, MARATHON, RUNNING CLUB, NIKETOWN RUNNING *CLUB, RUNNING EVENTS, RUNNING EVENTS SEARCH, BOWERMAN.*

SEARCH ENGINE DIRECTORY SUBMISSION

There are two types of search engines: those that use spiders to crawl your Web site and index it, as discussed above, and those that rely on human-powered directories to feed them their content. In order to appear in results from these search engines—search engine directories—, you need to submit your Web site to the directory and be approved. However, you don't submit your entire Web site, but rather a short synopsis of your Web site. Sometimes a good, professional looking Web site can be found by the directories themselves, independent of being submitted. In this case, the Web site is lucky and doesn't need to submit to be listed. Search engine submissions are used by search engines to help rank and register your Web site. A strong listing with directories can improve your chances of a top listing on many search engines.

LINKING PROGRAMS

Another criterion that search engines have typically used in their algorithms to determine where Web sites will appear in natural search listings is the number of relevant links that Web sites have linking to them. Consequently, reciprocal linking with other similarly themed Web sites will help drive relevant traffic to your Web site from those sites. Building relevant two-way links to your Web site is one of the most important things you can do to increase your organic or natural search engine rankings. Link popularity updating is one of the best ways to quantifiably and independently measure your Web site's online awareness and overall visibility.

Affiliate marketing programs or strategic alliances can also serve as great additions to your linking program while also generating incremental revenue. An affiliate-marketing program is defined as a system of advertising in which Web site A agrees to feature a button or link to Web site B, and Web site A gets a commission on any sales generated on Web site B from the traffic that came from Web site A. It can also be applied to situations in which an advertiser may be looking for marketing information, rather than a cash sale. A strategic alliance may not have the revenue component associated with it like affiliate-marketing, but entails reciprocal linking with your alliance partner for whatever the strategic reason you determined when you set it up. For instance, maybe you form a strategic alliance with a company who sells the same product as you, but serves a market or target audience that you physically or operationally cannot. An online alliance could definitely be beneficial to both of you from a traffic generation standpoint. Either way, whether it is an affiliate marketing program or a strategic alliance, you will likely get credit from the search engines for the associated links to your partners, which should help your SEO efforts while you are generating revenue and qualified traffic.

PRESS RELEASE OPTIMIZATION

Press release optimization entails the online distribution of press releases with newsworthy information about your company, products or services in an effort to receive online visibility among your target audience(s). By distributing your press release online via appropriate submissions, you can optimize keywords within the press release—much like you do for your Web site—and with the right keywords, effective optimization and distribution, press releases can garner significant, traceable traffic to your Web site and in your front door.

The news sections of search engine Web sites, such as Yahoo News, Google News and MSNBC Searchbot have become very popular and get millions of daily Web site visitors. Many journalists and news aficionados are using these search engine Web sites as well as other news aggregator services to find news and information. It's easier for them to do this now with the advent of RSS feeds (as mentioned earlier and discussed further in the section 4.5).

Like any online marketing tactics, press release optimization can be implemented and managed by you or someone within your company, or you can outsource it to a search professional. Either way you elect to go, here are a few reminders and tips to consider:

- ➲ Use online press releases to not only inform the public and the media of your newsworthy announcement, but to drive traffic to your Web site—make sure to include links within the press release to other relevant content on your Web site.
- ➲ If you do-it-yourself, use a press release distribution service. There are Web sites that offer free services and some a fee-for-service. The caution with the Web sites that offer free distribution services is to be careful that they aren't associated or accused of any press release spamming. Examples of fee-for-service Web sites include: PRNewswire.com , Businesswire.com, Webwire.com, or 24-7pressrelease.com. Examples of free Web sites include: PRweb.com, i-newswire.com, or PR.com.
- ➲ Use online press releases to increase your links and build link popularity.
- ➲ Make sure your online press releases are submitted to RSS feeds so that they can be served up on news aggregators or so individuals can subscribe to your feed.

➲ When authoring an online press release, try to use anywhere from 3-6 keywords or keyword phrases that associate with your press release content, your company or your products/services. It is believed that by inserting a keyword phrase at the beginning of a sentence, paragraph and in a heading or headline will help ensure your chances of getting picked up by a search engine.

➲ When developing your press release headline it doesn't need to be as catchy as traditional press release headline protocol dictates, but rather make sure it includes one or two relevant keywords.

In the end, the PR industry has evolved technologically and there is great opportunity these days for gaining exposure for your company, products or services through press release optimization. Journalists and the public (and your target prospects) are out there searching on the web—and they are likely searching for information within your industry, whether it be via search engines, news search engines, or news aggregators.

VIDEO SEARCH OPTIMIZATION

A note regarding the growing online video trend and optimizing online video for search engines: Search engines have developed technology that can search videos and produce links to them as part of their listing results. Through Really Simple Syndication (RSS) technology, video search optimization is becoming more and more popular and with the projections associated with the online video arena, we expect to only see this trend proliferate. Video search optimization relates to posting videos/video clips on your Web site or a social network (such as YouTube) and then making efforts to promote them so they appear in search engine results. We address RSS and online video more in later sections of this module.

REPORTING & TRACKING

Regardless of the type of SEM program that you elect to implement and whether you elect to do it yourself or hire an SEO professional to administer and manage the program(s) for you, it is highly recommended that you regularly track and manage your results and progress. If you hire a professional, at a minimum, request a monthly status report. SEM is generally not something that you can implement and walk-away. It's like a campfire, it takes constant attention and it needs to be fed, or it will go out. For instance, the competitive keyword landscape within the search engine could change, you wouldn't know it and the traffic to your Web site could plummet—thus, your fire would go out.

4.4 EMAIL MARKETING

Purpose: Understand email marketing as a tactical element and if you elect to include it in your marketing mix, understand the importance of implementing it correctly.

Key Points: In this section you will learn the proper way to set up and implement an email marketing tool and tips for effectively communicating with your subscribers or recipients.

Email marketing has legitimately become a staple of many marketing plans and if it isn't already part of most company plans, it's on their radar. As testimony, according to eMarketer, the total number of emails sent in 2007 is projected to be over 2.7 trillion while Jupiter Research has reported that 93% of executives surveyed in a recent study indicated that their company had implemented some form of an email marketing solution. Finally, the Direct Marketing Association has estimated that spending on commercial email marketing solutions in the United States will hit $600 million in 2008. All are compelling statistics that email marketing is big business now. It's gone from the neat, new tool, to a business must. Every professional utilizes email. Therefore, email marketing continues to be a cost efficient tool for promoting products and services while generating and retaining customers. In this section we cover email marketing tools as a tactical element to add to your online marketing mix and show how to effectively embark upon implementing and using one for your organization.

NO SPAM

Before you get started, first a quick note on what not to do. Simply, don't spam people! Spam is the inundating of the Internet with many email messages in an effort to send people email messages when they would otherwise elect not to receive them. The majority of spam is commercial in nature and the senders, or spammers, are frequently companies selling and promoting suspicious and/or questionable products and ponzi schemes.

Because of spammers, the United States Congress passed the Controlling the Assault of Non-Solicited Pornography and Marketing Act of 2003 (the "CAN-Spam Act" or "Act") on December 8, 2003. The CAN-Spam Act defines commercial email as any email message with "the primary purpose of which is the commercial advertisement or promotion of a commercial product or service."

The CAN-Spam Act creates a uniform "opt-out" standard, meaning that it prohibits the transmission of unsolicited commercial email unless the recipient has asked the sender to receive such messages. The Act also creates new criminal penalties for falsifying transmission and routing information in email headers or engaging in other practices designed to conceal the identity of the sender. Email header information that is technically correct nonetheless will be considered false and misleading under the Act if it includes an originating email address that the sender accessed through hacking or fraud. This provision is intended to address situations in which spammers obtain unauthorized access to an innocent party's email account and use it to send spam.

Spam is not to be confused with a newer email term, BACN (pronounced "bacon"). Bacn refers to any email you receive that feels as though it's cluttering your email inbox and sometimes gets annoying, but it's actually email you want to receive and eventually read. Granted, you probably wait to read it last. Examples of bacn email are: bank statement notifications, e-news alerts from companies, newsletters, coupons and/or promotions for products you asked for, etc. It's not spam, but it's not personal email either. Its email that was more than likely intended to market to you. The difference is, you asked to receive it at one time or another. Bacn is the type of email that we will talk about and advocate later in this section.

EMAIL STRATEGY AND PRACTICES

If you determine that email marketing is a tactic that will help you achieve your online marketing & sales strategy and objectives, we recommend that you adopt and adhere to some strict practices that will help you build and maintain goodwill with all your subscribers or recipients, increase your chances of keeping them as subscribers and probably ensure that you don't get any calls from the CAN-Spam police. We refer to it as the E3 Code of Conduct, which stands for Email, Etiquette and Ethics. The code is made up of the following five principles of conduct as it relates to using email as a marketing communication channel.

1) *Opt-in Only* – Only send e-mails to those who have opted-in, or those who have knowingly given you their e-mail address in some form or fashion (we talk more about your subscriber or recipient list later).
2) *Frequency Sensitivity* – Plan and maintain a communication stream for each of your opted-in subscribers (or lists) so you can manage the frequency with which you send messages. Some people may sign-up

for emails that you send as frequently as weekly, but for the most part a good rule of thumb is not to exceed more than 2 per month.

3) *Unsubscribe Ability* – Give each recipient the ability to unsubscribe at any time on any email. Even explain or offer a link to a page that explains your privacy policy and that you do not intend to use their email address for any other reason other than communication between the two of you.

4) *Relevant Content* – Do not abuse the permission level that a subscriber has granted you. Communicate clearly at the time you request someone's email address the type of information and content that you intend to send them and then stick to it.

5) *Tone* – Never forget who you are talking to. Remember that your recipients are customers or prospective customers. Speak to them in your emails in a tone and manner that you would speak to them in person. Never talk down to them and always maintain a polite and appreciative tone. Realize that your ability to communicate with them can be eliminated with the simple click of a mouse (i.e. if they choose to click on "Unsubscribe").

Once you decide to make email marketing a tactic within your online marketing mix, you will need to determine a tactical strategy. In other words, what is the email marketing tactic going to contribute towards the overall strategy and how are you going to execute or implement it? Will it purely be a promotional tactic? Or will it be more? This will obviously depend on your particular product, service or situation.

However, given the two-way communication potential that email offers, we recommend that you consider a tactical strategy that isn't simply pushing advertising or promotional messages out to your subscribers or recipients. We recognize that you likely want to use the medium to generate sales, but make the strategy a hybrid between promotion and relationship building. You want to try and do as much as possible to have your email messages perceived as "personal email", as opposed to bacn. Remember from the previous section—if you're bacn, you'll likely get read but maybe not immediately and if so, you might only get skimmed before they delete you. By contrast, personal messages get attention. What makes them get attention? There are usually two criteria: 1) it is from a trusted source with whom the recipient has a good rapport or relationship; and 2) the recipient is predisposed and conditioned to believe that the message will have more relevant information and something of value to them. If all you ever do is hit your recipients up to buy something, your message will grow stale quickly.

We advocate that you think about setting up your tactical strategy with the goal of gaining the equivalent of "personal email" status with your subscribers or recipients. To do this, you will need to make sure that you are generally:

➲ <u>Including content in your email messaging that is valuable to them</u>
➲ <u>Not too aggressive with your promotional message</u>
➲ <u>Asking them to take part in a dialogue with you</u>

Consider setting up an email calendar that maps out all the times you intend to send email in a given year—much like you would any other media calendar. If you have different emails going to different target groups/segments then create a calendar for each. Additionally, take it a step further and map out your messaging for the year. What message will each email contain? This will help ensure that you stay consistent—that the email marketing tool helps you achieve your overall online marketing & sales strategy objectives and that you don't send to your recipients too frequently. Sending too frequently could prompt recipients to unsubscribe.

EMAIL FORMATS

There are various email formats that you can utilize for your email marketing efforts. The format you elect to use will largely be driven by your email marketing strategy and objective(s) discussed earlier. An overall email marketing strategy may include different emails going out to constituents in one or more formats. Below is a list of some of the more common email formats, some of which you probably receive or have received in your own email inbox:

➲ E-Announcements
➲ E-Promotions/Catalogs
➲ E-Newsletters
➲ E-Zines
➲ E-Press Releases

CREATING THE LIST

The first step is to understand the true definition of an "opt-in." You should only send email messages to subscribers who have opted-in. By opt-in we mean people who have given you explicit permission to have their email address on file and to send them agreed upon information at an understood frequency. This definition does not include lists from list brokers who call you and explain that everyone on their list has opted-in. Don't fall for that. In fact, we advocate sending a confirmation email to your opt-in subscribers asking them to confirm their opt-in—this is sometimes referred to as a double opt-in. Double opt-ins mitigate computer programs that are designed and programmed to automatically find email sign-up forms from successfully registered fictitious names. They may get fictitious people signed-up, but when they don't confirm their opt-in, you know they weren't for real—or if it was a real person that didn't confirm, chances are they changed their mind. If you build your list to E3 code standards mentioned earlier, you'll know that the opt-in subscribers on your email list will be specific for your company or product. According to Jupiter Research, opt-in emails account for 27% of all email received in individual's inboxes, which is an increase of 11% since 2003. And with the growing emphasis on protecting privacy, this number is likely to continue to climb.

This means that you may be starting with very few people on your list (i.e. subscribers) and that you may need to add a sub-strategy to your tactical strategy and that is—a list growth strategy. It may not make sense to implement an email marketing tool until you have gathered a pre-determined number of email addresses.

We have a non-profit client who we have convince should evolve their printed newsletter into an e-newesletter. They have over 7,000 constituents in their database, yet only 592 email addresses from individuals who have said "Yes, you may email me." That's less than 10%. One of the main objectives for moving to an electronic newsletter is to save on printing and postage expenses. We realize that there will be a period of overlap where both a printed and electronic version of the newsletter will need to be distributed until critical mass is reached. However, 592 is not a significant enough quantity to justify doing both yet. Therefore, we recommended they shift their immediate attention to an intermediate strategy—a list growth strategy. We set an objective to grow the email list to reflect 20% of all constituents (1,400 email addresses) before implementing the first issue of their electronic newsletter.

Growing your opt-in email list should be a multi-faceted effort. Basically, to grow it as quickly as possible, you need to make it a company-wide effort and not just the marketing department's problem. Below are several examples of potential email address capture points:
- When prospective customers/customers call your company
- When sales representatives are on account calls/sales calls
- Building an online email capture form on your Web site
- Directing people to the online email capture form in your other marketing tactics
- Having opt-in/sign-up forms at trade-show booths
- Having opt-in/sign-up forms at the point-of-sale in retail locations

MESSAGING

As we mentioned in section 4.1, you'll want to keep your overall messaging in any emails consistent with your other marketing tactics (both traditional and online). Many email marketing tools, which we will talk about next, offer the ability to create HTML emails. This gives you the capability to extend your brand or product in a much more aesthetically pleasing and impactful way than just text. However, depending on the purpose of email as a marketing tactic for your company, you may not want it to appear too much like an advertisement or promotional piece. Regardless of the purpose, when it comes to authoring the copy for your emails, whether they be 100% text-based or HTML-based, there are some things to consider that will help you to not only improve deliverability (i.e. the chances of your email not being blocked by a spam filter), but readability. Below are ten writing tips for authoring effective email marketing messages:

1) Avoid the using large font.
2) Avoid the use of serif fonts.
3) Avoid using all caps.
4) Keep the use of bold or italics to a minimum.
5) Avoid using cliché words or spam target words.
6) Avoid fluff words, industry jargon or industry acronyms.
7) Avoid complex words and language.
8) Keep subject lines short but intriguing.
9) Keep paragraphs short.
10) Leverage links and blogs.

SELECTING AN EMAIL MARKETING TOOL

There are probably hundreds of different email tools or email marketing engines out there for you to choose from—all of which are online email/software applications such as Swiftpage, Constant Contact®, Exact Target®, Email Labs, or NetNewsdesk. An email marketing engine is an application that does more than simply create and push out emails. Your Microsoft Outlook can do that—well, it can't create an HTML email, but you can drop HTML images into your Outlook and send them. No, we mean a robust application that can not only help you graphically create your but that also has a database for your subscribers with the functionality to track open rates and other performance measurements. The email tool that you select will likely be a function of any one of the following. For each item, we offer some tips and/or things to look for as you investigate your options:

Budget

➲ This is often the biggest driver in a company's selection. You may naturally like many, if not all of the other criteria or factors listed below, but it may be that you simply cannot afford them. You will discover that there are large price variations in the email tools, ranging from practically free to very expensive systems with significant licensing fees. If you want to have the vendor or agency help you design, create and execute the emails and manage your email campaigns on an ongoing basis, those services will impact your budget as well. Most email tools are designed to be self-sufficient. However, be cautious when deciding whether to allocate budget to have someone help you. We've seen clients think that it will be easy to manage their own email tool and subsequent email campaigns—and set out to be self-sufficient, but when managed effectively and with the proper attention, it is more time consuming than many often think.

Frequency

➲ This one is simple. If you only plan to email occasionally (e.g. 6-10 times per year), it probably won't make sense for you to opt for an email tool that has all the bells and whistles. More importantly, we would want to speak with you about your intended frequency—if this is your choice. From a strategic standpoint, if you are considering incorporating email into your marketing mix, we recommend you plan to send at least monthly. Additionally, the email tools will charge you for each email that you send—usually on a per email basis. It is very important to know that the number of emails you plan to send in a given year will give you leverage to negotiate a lower cost per email with your service.

Subscribers

➲ How many subscribers do you intend to have and will there be a variation in subscribers? Do you plan to be able to collect varying demographic and psychographic information on your subscribers and segment them accordingly? The more you plan to know about your subscribers-their demographics, the products of yours that they use, their behaviors, etc.—all the things that will determine how you will speak to them to zero in on your calls to action—the more robust you system will need to be. Many email tools come with databases and profile centers, or profile engines that are really the hub of the system. They allow you to set-up the information you want to collect, collect it and subsequently mine and sort the data accordingly before sending any emails.

Deliverability

➲ Be sure to inquire with any application provider you are considering about their deliverability. They will each make some deliverability claim such as "We have a 98% deliverability rate". Be wary of the vendor who guarantees 100% deliverability. It's not possible. Also, ask them what their definition of delivery is. Is it from their server (where the email will originate) to the recipient's server, or is it to the recipient's desktop. A note of caution: Server to server numbers have the propensity to be misleading because they don't factor in any of your emails that get stopped by a recipient's company firewall or the spam filter on their desktop. Server to desktop deliverability rates are more accurate.

Future

➲ The vision you have for using email as an online marketing tactic both now and in the future will impact the tool you select. If you know that email is where you need to be and are confident that it will become a significant part of your online marketing mix, then you may want to anticipate and go with an email tool that has features that you may not use immediately (such as survey tools, dynamic content generation, etc.). This may cost you a little more now, even though you may be arguably paying for the features you aren't using, but it will be seamless to grow into those features. Otherwise, if you are uncertain about the power and potential of email marketing tools and you are suspicious of them, you can go with a less expensive option with fewer capabilities. Realize also that if you go with the less expensive email marketing tool initially, you can always upgrade to a more advanced tool later. Just realize that there may be some costs associated with switching brands/tools and/or switching inconveniences.

Access

➲ This may or may not be an issue with some email tools, but depending on how many team members that you expect to need access to the email tool—e.g. to either load data, analyze data, design emails, author content, coordinate an email send, etc.—the number will sometimes impact pricing. Some of the tools charge for access to the email tool on a per person basis.

Tracking

➲ Be sure to inquire with each email marketing engine about their tracking capabilities. This is a critical component to being able to measure performance metrics for your email campaign(s) and ultimately tie them back to your online marketing & sales strategy/objectives. For instance, if you have links to your online store, or a page with some sales related call-to-action, you will want to pay particular attention to them and depending on your company's ability to automate conversion tracking, you may need to or want to implement a manual process of analyzing those who clicked on these links in order to trace sales back to an email message and understand your conversion rate for the tactic.

4.5 SOCIAL MEDIA

Purpose: Understand the definition of social media including all the options and how to evaluate its impact on your business.

Key Points: In this section you will learn a process for evaluating and implementing the correct social media tools. Just because it exists or is the new "buzz technology" does not mean you should use it. You will understand how to locate research, evaluate options, and prepare for implementation.

Have you been socially networked yet? If you haven't participated yet be patient, wait, and you will. Recently, the web has created large social networks where the users generate the content. According to SiteProNews, "This phenomenon has given consumers a voice and weakened the power formerly held by advertising media. Social media, therefore, becomes increasingly important to a Web site's success and its visibility in search engines".

As you begin to understand this new frontier on the web, make sure you understand the ramifications of participating or not participating. According to Michael O'Neal in ITBusinessedge, "Meaningful debate will blossom only within communities that are passionate about issues. These communities will include those defined simply by a desire to discuss/debate".

Social media is a big buzzword within the online marketing space today. It is essentially synonymous with Web 2.0. By definition, social media refers to online mediums where the users are participants in the publication of and exchange of content (such as text, news, photos, videos, and audio, etc.). Social mediums include such things like: forums, message boards, blogs, podcasts and social networks—all things that allow for the exchange of information collaboratively. In this section we will specifically discuss blogs, podcasts and social networks and how they can be leveraged as an online marketing tactic.

There are tons of new sites coming online every day to help you network or connect to others. We are witnessing the snowstorm before the avalanche. If you're already doing it, then great! If you're not, here are a few do's and don'ts:

➲ Do join social networks that fit you and your tastes.
➲ Don't overindulge on your profile because no one cares if you were citizen of the month in sixth grade.
➲ Do get involved with the discussions and forums when time permits.

- ➲ Don't live on the network-people will wonder what you really do for a living.
- ➲ Don't blatantly advertise your offering but rather weave it in with good content.
- ➲ Do understand what your goals are for being involved.

BLOGS

The word blog is short for weblog, or broken down, it is a web log. Thus, it is a log or journal of web postings that are made on various subjects and then logged for historical reference so all involved in the conversation can see what is or has been said. It is literally a log of messages exchanged over the web between potentially unlimited numbers of participants. There are also other various blog formats, such as video blogs and photo blogs. Blogs offer the ability for the blog owner to categorize and post content to a web page and allow visitors to reply, comment or provide feedback on the posting at the owner's discretion. As postings are added over time, the previous postings are logged for reference at a later date, if desired.

Blogs exploded in popularity between the middle of 2005 and 2007. As of March of 2007, Technorati, a recognized authority on blogs, reported the existence of over 70 million blogs with 120,000 blogs being added to the blogosphere everyday—that's 1.4 blogs being added per second. There are about 1.5 million posts to blogs made on a daily basis, or about 17 posts every second. As of the writing of this book, Technorati reports that it is tracking 112.8 million blogs in existence.

BLOG PURPOSES

In addition to the different formats mentioned above (i.e. text blogs, photo blogs, audio blogs and video blogs—all now exist), for each of those, if you were to take inventory of the over 100 million blogs in existence today, you would find that there are blogs for virtually almost any topic from personal blogs, such as individual travel diaries, to very successful business-related or industry blogs (i.e. health technology), to special interest or hobby blogs, such as entertainment, shopping, sports, home & garden, to internal vs. external blogs (i.e. blogs whose audience is an internal audience over an Intranet vs. an external audience of customers, etc.). In addition to text blogs, there are photo blogs, audio blogs and video blogs. Any blog will likely fit into multiple categories depending on how they are being categorized. In every case, their intent is to capture the power of the medium to converse and develop rapport with others. For the purposes of this book, we will look at blogs from a business perspective focusing on capturing the interest of customers and prospects to ultimately build brands. Blogs can be effective marketing and communication tools for each of the following purposes (*Note: while there are likely many other purposes, they more than likely can be rationalized to fall under one of these reasons*):

➲ Information Sharing & Relationship Building
➲ Promotion & Public Relations
➲ Product Information/Education/Development
➲ Customer Service

IMPLEMENTING A BLOG

So given that you now know how blogs can potentially contribute to a business's ability to promote itself and/or communicate with its constituents, if the purposes described above match up with any component of your online marketing & sales strategy and corresponding objectives, then blogging could be a good tactic to consider as part of your implementation plan. You will, of course, need to determine the blog format that is most suited for your business, as well as which of your target audiences you will cater to and subsequently what your blog's purpose(s) will be (as described above, realizing that it can strive to have more than one purpose). You will also need to determine how you will create your blog. Web development firms can create them, but there are also a number of very robust and capable online blog platforms on the market that enable easy set-up and maintenance and have great tracking tools and features. Some of the more popular platforms include: Typepad, Blogger and Wordpress.

Here are some other helpful hints relative to developing a blog and becoming a member of the blogosphere. Note: this list is by no means an exhaustive list of what to do, but should give you a good idea of what you need to do and some pitfalls to avoid:

Communicate internally/encourage participation
➲ Be sure to communicate internally within the company that you are about to launch a blog and why you are going to do it (again, what's the purpose). Encourage everyone to visit the blog, subscribe to its RSS feed so they are notified of new postings and comment on postings. Also have them help spread the word that the blog is out there to encourage additional traffic to it—you'll be amazed at the power of word-of-mouth promotion and how quickly it spreads (assuming it's a relevant and entertaining blog). Finally, set a blog policy relative to posting and commenting on the web. Remind your team members that while you want them to express their opinions

honestly, the web is a public domain and there are ramifications to inappropriate conduct when blogging—negative press for the company and consequences to the responsible party.

Select a name

➲ In addition to the URL that you will select within your blog platform software, you may want to consider getting your own domain name for your blog as well—something that you can simply redirect to your URL from the blog platform. For instance, we have a blog for this book called webplicityblog.com that redirects to webplicityblog. typepad.com. Here are also some things to consider when selecting your name: make sure it associates with your company and brand in some way; try to name it something that might help describe the content of your blog; try to include a keyword(s) that you might want to optimize; and keep it as short as possible.

Identify content and authors

➲ Identify a structure for your blog that includes fun and relevant content that will be perceived as entertaining by your target audience. Avoid the temptation to make it one big advertisement for your company, its products, or its services—that will turn people off. Try to think of topics that are interesting to those in your industry. For example, if there are legislative initiatives that are constantly impacting your industry, talk about these. State your opinion about them and ask others to join the conversation. What other topics can you think of that would resonate with your readers, are hot within your business or industry and would spark conversation and debate? These are the topics that will make people subscribe to your blog. They'll always be curious as to what is going to be posted next. Concurrently, if they ever need a product or service within your industry, who do you think is going to be top-of-mind with them? Also, identify who will be the author(s) of the blog postings on your blog. Will it be one person or will you have multiple authors? The blog software platform that you select will allow you to set it up for more or many—however, you will need to identify an admin that controls who is permitted to author and who isn't.

Posting frequency

➲ There is debate within the blogosphere about the frequency with which to post to your blog. Some advocate that you need to post frequently in order to get noticed, while others say that posting too frequently can impact content quality. Our recommendation is to post when you have something to say that you think is relevant, worthwhile and valuable to your targeted readers. That said, if you're going to commit to blogging, be willing to commit to searching out content and authoring postings. Like athletes, blogs need food to perform. Content is their food. If you don't feed them, they won't serve their purpose and you won't achieve your objectives.

<u>Be authentic and real</u>

➲ The key to blog postings is honesty, authenticity and genuineness. The blogosphere can sense phonies and fakes. There are legendary stories of companies who hired someone whose pen name was subsequently the blog author. It didn't take long for the blogosphere to figure this out and the blog was essentially abandoned—people stopped visiting. Its integrity had been compromised. Strive to speak from experience and with emotion. Avoid business talk, techno-talk and company jargon. Speak the truth and what's on your mind, not what you think your audience wants to hear.

<u>Monitor posts, respond</u>

➲ The blog platform software that you select will have many features associated with your postings. One of the key features entails the ability to allow or disallow comments for any given post. You can indicate at the time you make a post whether you would like to allow the functionality for people to comment on it. Use this tool. For example, if there is something you are going to post that may be especially controversial—to the point that it could elicit negative press for your company—you may want to disallow comments. For postings where you do allow comments, you can set the tool to allow you to approve all postings before they are made public. When you do allow comments, whether you are approving them or not, monitor them and when necessary or if appropriate make sure you respond. Your audience needs to know you're listening (in this case reading).

<u>Monitor blog traffic and statistics</u>

➲ Be sure to monitor your blog traffic and other statistics such as length of user sessions, number of new visitors, etc. Traffic will help you justify whether your blog is working or not, or whether it's worth continuing to invest company resources and time. Like email, you can include links to your online store, or a sales related call-to-action, but this tactic may not be the one that drives the most sales. Its mission may be more to build and foster relationships while some other tactic might be the closer. But if traffic continues to improve month-over-month and you're hearing anecdotal feedback from your sales force, or blog authors that people love your blog—and other comments like that, those may be all the metrics you need.

Protect yourself

➲ Consult your attorney or legal counsel and make sure you develop some legal disclaimer text to include on your blog that protects you as much as possible from any liability. Something to the effect that other than the original postings, you monitor comments that are posted and attempt to keep inappropriate responses from being displayed. However, sometimes they may be missed, or opinions are posted that are not necessarily inappropriate, yet the opinion expressed may not be shared by the company. A posted response or comment does not constitute an endorsement of that viewpoint by the company.

PODCASTS

The root of the word "podcast" refers to an iPod, which is an Apple Computer product. However, you don't need an iPod to podcast; iPod is simply the namesake for the medium. Podcasting is the act of taking a digital audio file, like an MP3 and distributing it to registered listeners via Really Simple Syndication (RSS), or an RSS feed. When a listener has registered to receive a podcast through an RSS feed, the podcast is automatically downloaded into their media player (i.e. iTunes, Windows Media Player, etc.). Evolving right behind podcasting is video podcasting, which is beginning to be talked about more and more since technology will now allow for the distribution of a video file over an RSS feed. The only difference as of right now is in how podcast and video podcasts are consumed by listeners/viewers. Podcasts can be delivered to a mobile device and listened to at anytime—even while driving. By contrast, while a video podcast has the same capability, as long as the device you are using has the ability to view the video, it is not suited as much for mobile devices, absent of watching it on an iPod while on an airplane or something like that. Thus, video podcasting tends to be more of a desktop consumed medium right now. That said, it is still so new we couldn't even find any measurements relative to its market penetration.

While the awareness for podcasting has jumped, reportedly from 22% to 37% (68% increase) as reported by Edison Media Research, it is still a very new medium. The same Edison Media Research report indicates that those who reported listening to a podcast grew from 11% in 2006 to 13% in 2007. Additionally, earlier this year eMarketer reported that the 2006 active podcast audience (or population) was 3 million people and the potential audience was 10 million people—with *active audience* members defined as individuals who download one or more podcasts per week and *potential audience* members defined as individuals who have never downloaded a podcast. They forecast podcasting audiences to grow to 7.5 million (active) and 25 million (potential) members by 2008 and 18 million (active) and 55 million (potential) members by 2011.

NICHE TACTIC

So are podcasts right for you? Does it make sense to pursue podcasting as a tactic for your online marketing plan? Given the aforementioned statistics and the medium's projected growth curve over the next four years (e.g. it will only attract an audience of 18 million, or approximately 6% of the U.S. population) it is definitely a niche tactic. Therefore, determining whether podcasting is for you is going to be predicated on one or more of the following four factors:

1) Target Audience

Is your target audience tech savvy? Are they innovators in that they are always on the forefront of new technologies, ideas and equipment, etc.? If so, then podcasting might be an effective means with which to reach them. Plus, it may help them to perceive your brand as an innovative and forward thinking brand.

2) Products and/or Services

Does the nature of your products, services or industry in general lend to the need to be associated with such a non-mainstream medium? Will it help establish the image you need? Have you already run across information on similar products and services in an audio format? Have you heard of or seen your industry featured in an audio format? If you can answer affirmatively to either question, can you envision talking about your company in an entertaining fashion that will make people want to listen and subscribe? If so, then podcasting might be a great way to showcase your wares.

3) Brand/Product Image

Do you need your brand or image to be associated with a leading-edge, hi-tech medium such as podcasting? What will being able to say that you podcast or being able to tell prospective customers to go out to the web and sign-up for your podcast do for your brand image? Will it establish credibility? Will it establish you as a thought-leader and subsequently the best choice for your product or service? If so, then podcasting might help meet a brand building, awareness generating objective that is somewhere in your online marketing & sales strategy.

4) Budget

If after considering all of the above factors, you determine that the medium still doesn't have a significant reach to your customer profile and that podcasting is still too new, this option isn't for you. If the opposite is true and you've identified that your target audience is largely contained within the 3 – 7.5 million people that make up the potential podcast audience, then you should seriously consider podcasting as part of your overall marketing & sales strategy.

IMPLEMENTING A PODCAST

If one or more of the factors above make sense for you, then you have one of three options to pursue in implementing podcasting: 1) create, direct and produce the podcasts yourself, 2) have a professional create, direct and produce your podcasts, or 3) you create your podcasts and have a professional help you direct and produce it.

1) Create, direct and produce it yourself

Creating and producing your podcast on your own will require the following: time, some technical knowhow and some recording equipment and editing software that can actually be purchased at very reasonable prices.

2) Have a professional create, direct and produce

This is the easiest option for you, naturally. This will still require your time, especially if you are going to star in your own podcast, or if someone within your organization is going to be the voice talent. Also, it will require some planning meetings to approve the structure of the recording and how it will flow, etc. Afterwards, the professional edits the podcast (identifies the good takes, etc.) and comes back to you with a finished product. Note: As one might anticipate, this is the most expensive alternative.

3) Create it yourself and have a professional direct and produce

This is obviously a hybrid of the two aforementioned options. This option will give you creative freedom and arguably save money that you would otherwise pay the professional to help you concept the podcast structure and content (topics, discussion points, questions, etc.).

Here are a few helpful hints relative to creating, directing and producing a podcast. As with the list of blog implementation ideas above, this list is by no means an exhaustive list of what to do, but should give you a good idea of what you need to do and some pitfalls to avoid:

- Listen to other blogs for good ideas – what to do and what not to do
- Identify a format that is right for you
- Prepare for your podcast and make sure your guests are prepared to be recorded
- Be real and genuine
- Find a conducive environment for recording
- Pay attention to your speed and tone
- Don't forget a call-to-action
- Don't make it longer than the time span your audience will give you to listen

SOCIAL NETWORKS

Online and social networks have gained significant popularity in a very short period of time. The MySpaces, YouTubes and Linked-Ins of the world, just to name a few, have gone from start-up social communities to big business. This is evidenced by MySpace's acquisition for $500 million by Rupert Murdoch and YouTube's acquisition by Google for $1.65 billion. These acquirers see the marketing potential that these social networks possess in terms of captive audiences and track-able access to their audiences—or potential customers for other marketers. In late 2006, eMarketer estimated that ad spending on social networks would increase to $260 million in 2007, up from total spending of approximately $95 million in 2006. They also predict that ad spending will grow to $2 billion by 2010, which represents 485% growth over a 3 year period or an average of 161% per year. And a separate report by Datamonitor predicts that social networks will peak in popularity on a worldwide basis in 2009 and plateau in 2012. Regardless of the predictions, the fact is that between now and then marketers will be continuing to spend and will continue to try to figure out how to maximize this relatively new technology and medium.

Right now, from a marketing perspective, social networks are being leveraged in one of two ways:

1) Paid advertising/ banner advertising
   Relative to the second way (below), this is the most traditional option and is really no different than purchasing paid advertising/links on a search engine and/or placing banner advertising on any Web site, which we spoke about earlier in this module.

2) Viral marketing
   Viral marketing is the 21st century term for word-of-mouth marketing. We should equate it to word-of-mouth marketing that is turbo-charged. It is the act of a marketing message spreading rapidly through

a community or among an audience by virtue of the message being forwarded by each recipient to their sphere of contacts.

For the purposes of this book, we will primarily focus our discussion on leveraging social networks to promote companies, brands, products or services through viral marketing efforts or online video.

VIRAL MARKETING

The name viral marketing is literally in reference to the *spreading* of a message. Much like a flu virus can *spread* rapidly and affect people almost exponentially overnight, viral marketing works much the same way. A few people are exposed to a message and in the case of social networks, they quickly sneeze the message to hundreds, thousands, if not millions of people simply by sharing it with those in their community (who then share it with their community and so on).

Viral marketing can be very difficult to plan or control. Again, much like the flu virus, sometimes it spreads in almost pandemic proportions, while other times you hear there was a flu bug that went through town two weeks after it's gone and wonder why you never caught it. The biggest challenge for marketers as it relates to viral marketing is finding or developing that viral marketing idea that is contagious. Often the best viral marketing strategy, in addition to trying to force the viral message, is to simply be great observers and monitors of what is going on within the social network space. Sometimes a community member within a social network will create and post something that relates to, features or includes your brand or product. As long as the posting doesn't position you negatively (if it does there's a whole other separate strategy you would need to pursue for that) and it begins to gain some viral momentum, there is opportunity for you to leverage the posting to your company's benefit. While it's the hot topic within the social network space, you need to act quickly and promote it through other mediums—e.g. email or your blog—tactics that will help spread the virus even more. However, you still have to monitor it carefully and make sure the message doesn't turn negative as the virus continues to spread.

An early example of great viral marketing before the days of social networks was that of MasterCard and their priceless campaign. This was the campaign where they went through a series of items and listed their price, but the final result was subsequently "priceless". MasterCard inferred that the priceless moments were often made possible by their credit card. After a few great TV commercials in this format, emails began sprouting up where people had made their own "Priceless" parity using some embarrassing moment of their own, or at the expense of some friend, etc. MasterCard picked up on this phenomenon and leveraged the popularity by creating a "Priceless" sweepstakes and even a "Priceless" contest where people submitted their priceless pictures (priceless.com).

Social networks have only helped viral marketing become more powerful and essentially a channel for leveraging viral messages. The critical characteristics of an effective viral marketing message are:

- ➲ Entertaining – the message has to provide some entertainment or no one will want to share it and spread the message.
- ➲ Authentic – it has to be real and genuine. You can try to stage something to appear authentic, but the social network community may see through it. However, if it's still entertaining enough then the community might forgive you for the lack of authenticity.
- ➲ Unselfish – don't try and push the marketing message and call-to-action too much. That isn't your objective. You should simply be glad that people are interacting with your company, brand or product and that you are making a good impression (all the more important that the viral message is positive, or it can have an entirely reverse effect).
- ➲ Awe-Factor – messages or content that leaves people in awe tends to garner more interest and impel people to share it with their friends and family. If it's possible, integrate this kind of content into your message.

A modern day example of a viral marketing effort that has seen some success is for the maker of the chewy mints called Mentos. Mentos observed that someone had posted a video on the web that was essentially a fun documentation of how a bottle of cola can explode when Mentos are put inside. They have since leveraged this on their Web site with links to all the videos that are out there demonstrating this experiment and even provide instructions for how to make your own Mentos geyser (u.s.mentos.com).

ONLINE VIDEO

Online video has grown rapidly over the past two years—essentially since YouTube received its first $5 million in funding in late 2005. It's literally been all uphill since then as it relates to growth of not just YouTube, but online video as a medium. While YouTube and other social network video Web

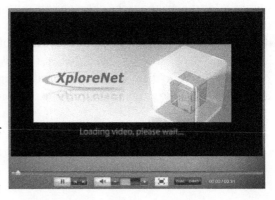

sites have begun to see more and more professionally produced online videos for big advertisers and the like (i.e. Cadbury and Smirnoff from the Side Bar above), the medium's popularity and growth can largely be attributed to user-generated content, or videos.

In 2008 and beyond we are definitely going to see this medium grow, evidenced by the big four broadcasters entry into the space. For advertisers the performance metrics of the medium are still not fully known, or at least not fully tested and proven. For instance, according to the Online Publishers Association even at this early stage in online video's life cycle, 31% of consumers who watched an online video (i.e. an online video ad), afterwards visited the advertiser's Web site. Additionally, 16% indicated that they mentioned the ad and the product in the online video to family and friends, and 15% reported that they were more motivated to go see the product at the store.

It's only going to explode from here. Again, with the major broadcasters delivering their programming via the web, who knows where it will go and how fast—but if we had to guess, it's going to be pretty quick, so fasten your seat belts. Broadband penetration will continue to grow and the advertisers who have been the early adaptors for the medium will continue to get more and more creative with how they approach the medium while the "me too" advertisers will all jump onboard (see *Online Video Watchers -- US Growth Projections*). The PEW Internet & Lifestyle Project recently reported that 62% of those who watch online video indicated that their favorite videos are those that were professional created. Because of this, we'll see more and more professionally produced online video content and probably less user-generated content. We predict that there will be a combination of both styles and that we'll see a lot of professionally produced videos that were purposely created to look user-generated. We'll also see advertisers test this theory and use the results in increasingly different and creative ways.

Earlier we focused on online video's use to date as a viral marketing tactic. It is smart to look beyond the ways that video is being utilized currently because the applications are endless. For instance, Carnival Cruise Line recently leveraged online video on a dedicated Web site solely created to give visitors a virtual tour of the cruise ship. It is literally a choose-your-own-adventure type video that stops occasionally until you tell it where you want to go next. It basically brings the ship to life on-screen. Any method that has been utilized in the past in person with customers will likely be adapted to online video in the near future.

Whether you want to develop online video content for viral marketing messages, or to communicate more traditional marketing messages, you will need to consider the following:

1) Understand your target audience and determine the quality standard that they expect for the video content. Can you get away with producing a piece that looks as though it was user-generated, or do you need to have it produced professionally? Obviously professionally produced video has higher cost implications than a user-generated format.

2) How do you want to use online video and how will you promote that it exists? Will it live on your Web site only, or will it make sense to post it and promote it among social networks? Will you video search optimize it (from section 4.3), or push the message via your own email efforts (from section 4.4)?

Bottom line-Social Networks can be a great medium or channel for disseminating viral marketing messages. For the time being, online videos seem to be the carrot for those messages that are making their way through the web. However, while social networks and online video will both continue to see significant growth, we can most certainly count on people adapting and evolving how they both are used. But for now, viral marketing and/or online video as tactics, if appropriate for your brand or product, has the potential to be a great addition to your online marketing & sales strategy Web plan.

Figure 4.6: Online Video Watchers – U.S. Growth Projections

According to eMarketer, below is a breakdown of individuals using online video shown as a percent of those who do so with broadband, those who use the Internet and the total population.

Year	Broadband Users	Internet Users	Total Population
2006	85.7%	62.8%	40.0%
2007	86.4%	72.0%	47.0%
2008	88.3%	80.0%	53.3%
2009	89.7%	84.2%	57.4%
2010	90.4%	85.4%	59.4%
2011	91.4%	86.6%	61.2%

Notes:
Online video watchers defined as people age 3 or older and who download or stream video at least once a month.
Population numbers source: U.S. Census Bureau.

Source: eMarketer, February 2007.

"This is revolutionary". says Jeffrey Cole of the Center for the Digital Future at the University of Southern California. Video killed the radio star but it's only enhancing the new web experience. Youtube, myspace, and other Web sites developed to allow users to post and download videos has led to an explosion in new technology and methods for web based video.

YouTube soared to 25.5 million visitors in November 2007, up from 900,000 a year ago, according to comScore. That's a 2,720% change. So the question is how do you in your role at your firm take advantage of this explosion especially when you want professional web content and not someone's drunk party pictures?

The reality is that you can promote your products and services through web based video professionally, respectfully, and profitably. Corporate video for commercials and training have been used for years with a great deal of success. Therefore, during the Drive phase of your web strategies you can learn about the current video options, importance of quality production, and implementation models that work.

VIDEO PLAYER

In order for a video to be viewable on the web the contents of the files must be compressed into a standard web video format to run on a player. Examples of this format include QuickTime built by Apple, Windows Media built by Microsoft, and RealPlayer built by RealNetworks.

The video content you're evaluating to put on the web must be high quality. The teenagers posting unclear, static laden video on youtube.com get away with poor quality, but your firm's brand will suffer if your content is not high quality.

4.6 AFFILIATE MARKETING & STRATEGIC ALLIANCES

Purpose: **Understand what an affiliate program and web based strategic alliance are and how to correctly implement them.**

Key Points: **In this section you will learn what an affiliate marketing process is and the best method for implementing it as well as the proper ways to form strategic alliances on the Web. We will cover some tips on potential vendors and processes and what to look for and how to implement the relationship. You'll understand how to maintain the alliances and keep them beneficial for all involved parties.**

Affiliate marketing has been on the Web for a very long time, but recently it has begun to make more sense than ever. An affiliate marketing program is defined in several ways. A system of advertising in which site A agrees to feature buttons from site B, and site A gets a commission on any sales generated from site B. It can also be applied to situations in which an advertiser may be looking for marketing information, rather than a cash sale. Also, it is a business relationship with a merchant or other service provider who allows your business to link to that business.

THE PROCESS

When a visitor clicks on the link at your site and subsequently makes a purchase from the merchant, you receive a commission based on the amount of the sale, a referral fee or a pay-per-click fee. These vendors typically sell goods and services or provide appointments or leads to you. You typically link directly to them and pay for either traffic to your site or the actual conversion of a customer. The program is based around password-protected technology that tracks the traffic or conversions. Most programs are set up for large retailers like Wal Mart, Target, or Dell. However, there are medium and small-sized programs that offer a low set-up fee and modest commission plans. You can easily target your established user groups.

Choosing the right affiliate marketing program on the Web to help you grow revenue is the starting point. Now, how do you implement a program? Here are some helpful tips:

➲ Establish your target audience.
➲ Set up the budget (what can you afford to pay to receive the customer)?
➲ Research the various opportunities on the Web.

Affiliate marketing is a Web tool that is often overlooked. Most firms are not aware of the benefits of implementation, nor do they know where to find the right program. You should spend some time researching the options and implement the program that fits your needs best.

STRATEGIC ALLIANCES

Every sales book in the world seems to describe the importance of strategic relationships. However, when it comes to the Web, many businesses either overlook strategic alliances or do not understand how to set them up. We touch on a few ideas and review a method to forming the alliances and keeping them strong.

Most businesses have strategic alliances, which are defined as agreements between firms in which each commits resources to achieve a common set of objectives. Firms often form strategic alliances with a wide variety of partners: customers, vendors, competitors, education sources, or government.

Through strategic alliances, companies can improve competitive positioning, gain entry to new markets, supplement critical skills, and share the risk or cost of major development projects.

USING THE WEB TO ALLIGN

How do we do this on the Web? The process includes the following:

➲ Define your business vision and strategy in order to understand how an alliance fits your objectives.

➲ Evaluate and select potential partners based on the ability of the firms to work together toward a common goal.

➲ Develop a working relationship and mutual recognition of opportunities with the prospective alliance(s).

➲ Negotiate and implement a basic or formal agreement that includes systems to monitor performance.

Strategic alliances for the Web typically include link exchanges and traffic push. This is where you incorporate a linking strategy to increase traffic to your site, and your results, in the search directories. When you have identified alliances, you can exchange Web site or application links to link users back and forth between the sites. In addition, you can push traffic to each other, specifying service offerings through content, text, graphics, and direct requests to visit the ally's site.

BENEFITS

The benefits of a properly established strategic alliance are immense. These benefits allow you to:

➲ Reduce costs through economies of scale

➲ Increase access to new technologies

➲ Increase knowledge

➲ Enter new markets

➲ Beat competitors

➲ Reduce sales cycles

➲ Increase traffic

➲ Improve your bottom-line

4.7 SALES DEPARTMENT/CUSTOMER RESOURCES
Sales Department/Customer Resources

Purpose: Demonstrate the value and benefit that the web can bring to your customers as it relates to selling to them or servicing.

Key Points: In this section you will learn some effective online sales tools and customer resources and tips for implementing them should you determine they are right for your online marketing & sales strategy/plan.

The web and subsequently your Web site have the potential to be a powerful sales and/or customer resource tool. You, or your prospects and customers can do so much more on the web today than you could back in the late 1990s even. The proliferation and penetration of broadband enables users to view large files—such as graphic intensive presentations, online video and other dynamic content quickly and with good visual quality. According to Internet World Stats, the U.S. has the most broadband subscribers in the World today. Over 66 million people have broadband, or 21.9% of the population. When you filter that down to just adults, a study by the Pew Internet and American Life Project reported claims that the % jumps to 47%. I repeat, forty-seven percent of the adult population in the U.S. has a broadband connection.

So what does this mean as it relates to online sales department tools or online customer resources as a component of your online marketing & sales strategy and implementation plan? It means many things. It means your sales department now has a reach that it never use to have—their territory is now as large as they want to make it. It means that the barriers to sharing sales information (i.e. demos, presentations, etc.)—things that used to require face-to-face visits—have been removed. It means that the ability to service customers and share information (even proprietary information) can now be done securely via long-distance in a matter of seconds. The list could go on and on.

Selling and servicing customers on the web is very different from doing it face-to-face. We aren't advocating that you replace live meetings with online sales tools or customer resources, but rather use them to augment your sales process or customer service when appropriate. The difficulties when it comes to using the web for these types of activities range from not being able to read body language to the inability to ask direct questions. However, leveraged effectively and with the proper forethought, they can be of great assistance, add value and even save time and money. How you can optimize the web and your Web site is dependent on your internal processes and is only limited to your creativity. Here are some online sales department tool and customer resource ideas and tips to get your wheels turning:

Web Presentations
➲ Use self-paced Web presentations to increase new customer sales through a "real" 24-hour/7-day a week, online store. Your prospective customers can go through presentations at their own pace and ask specific questions regarding different parts of your product or service. When streamed from your Web site, with a broadband connection, these presentations can be more impactful than ever—with the utilization of high-resolution graphics, video, audio, etc.

Optimize Sales Literature
➲ Save on costly printing and mailing costs by directing requests for product or service information to your Web site. Create brochures, product literature, service benefits and other sales documents in PDF (Adobe's *p*ortable *d*ocument *f*ormat). Make sure the content is clear and of high quality.

Online Audio/Video
➲ Include customer testimonials and usage demos in audio and video format in the sales and or product section of your Web site. You can create downloadable or streaming audio or video (make sure you offer multiple choices to your prospects, depending on their Internet connection).

Chat Rooms
➲ Enable prospective customers to chat with your sales personnel, either through a live chat tool or a discussion board. Prospective customers can post questions and sales representatives can answer them at any time of the day, including evenings, weekends and holidays.

Dynamic FAQ Sections
➲ Create an area on your Web site for frequently asked, sales-related questions that is updated automatically. For example, if a prospective client asks how many colors item 334 comes in, the question and the response can be automatically updated to the FAQ area. Therefore, when the next prospective customer comes into your Web site with that specific question, the question is already in there with the answer. Your sales team will look like heroes.

Special Programs Section
➲ Create a special programs section within your Web site, such as a frequent buyer clubs, giveaways, coupons, prizes, awards, etc. These

windows should pop-up or link directly from relevant product sections or the sales section of your Web site. If your traffic reporting software (see 4.10) allows it, have these programs, offers and promotions appear dynamically based on the profile of the visitor/customer.

Customer Service/Client Management Portals

➲ Create a login and password protected area on your Web site that clients can login to in order to see information related to their account, or see other valuable information, such as the ability to: communicate electronically with a company representative, pay and track invoices, review their account history, see sneak previews of new product releases, find special offers on product upgrades, etc. In time, aside from their account manager or sales representative, customers will perceive the portal to be their lifeline to your company.

Site Search Feature

➲ We have all become very accustomed to searching for information online via search engines like Google and Yahoo! We do so by typing in search terms relative to what we would like to find. Build this same functionality into your Web site so a visitor can type in specifically what they are interested in and your Web site will then take them there directly. Granted, if your Web site has been built with a proper Web plan and has an intuitive navigational structure, many visitors won't need this feature, but it's customer-friendly to have it available.

Display Pages Properly

➲ Try not to exceed 30-40 kilobytes of space for a single page when all the photos/graphics are loaded. If you need a larger photo size to see detail, make sure a "clickable" thumbnail is incorporated into the design. The thumbnail concept gives the customer the option to view the larger size for more detail. Detailed, quality product photos take up large amounts of disk space. Make sure your hosting company has the space for you. If they set your Web site up with 50 MB (megabytes) then make sure all MBs are accounted for. Lack of disc space will kill great photos and downloadable video clips quicker than bad design. If the photos and graphics don't load quickly enough because they're too heavy or because you don't have enough disk space, you'll lose customers – guaranteed. This isn't as critical as it was in the days before broadband connection speeds, but you should still be cognizant of your page sizes to ensure a quick loading Web site and subsequently a delay-free user experience for visitors.

Take Advantage of 3rd Party Conferencing Tools

➲ Utilize 3rd party web tools like webex.com, gotomeeting.com, kolabora. com, etc., to improve the sales process. These Web sites help users do real-time presentations, combined with conference calls and online chat meetings. When you can't have a live, in-person meeting due to long distance, or you need to conduct a quick meeting from across town, these applications can be far more beneficial than a just a conference call/phone connection.

Offer Email Alerts/Newsletters

➲ Allow customers to sign up for email alerts and/or newsletters regarding new products/services. These alerts should offer value as along with a sales pitch (see 4.4).

The challenge with selling and servicing customers via your Web site is that customers and prospective customers have the control, rather than the company or a company representative. The visitors determine what they see and when they see it. However, you can increase your close-ratio or improve your service-levels by anticipating their needs and offering innovative tools that deliver informative, impactful and desired information when they need it.

4.8 TRACK & ADJUST

Purpose: **Evaluate different tracking tools and processes and implement the proper changes.**

Key Points: **In this section you will learn to track your promotional ideas and make the proper adjustments. This includes tools, time frames, and objective data analysis. In the end, you will know when to stop a campaign or heighten the activity.**

Incorporating web traffic tracking or a web analytics tool into your Web site will likely be a critical component to enabling the tracking and measuring of your progress toward the objectives that we discussed in section 4.0. Again we recommend you incorporate a tracking option as part of your online marketing & sales strategy/plan. As testimony to the importance of web analytics, the market today is over $400 million and is projected to grow to over $650 million by 2010. Thus, by virtue of

the size and projected growth, many firms, including your competitors, are expected to increasingly come to realize its value and importance.

The number of web traffic tracking tool choices is somewhere in the neighborhood of 500. So the question becomes, how do you pick the one(s) that are right for you? At a minimum you need a basic software application to track the simple traffic on your Web site. Webtrends essentially owns the market with nearly ninety percent (90%) of web hosting firms offering a Webtrends package. WebTrends monitors and reports on traffic results for things like hits, page views, user session length, total visitors, unique visitors and channels.

TRACKING TOOLS

There is no shortage of reliable web metrics services and software available today. Unlike many of the early web metrics tools, those available today are more intuitive and user-friendly. By incorporating and learning to use a web analytics tool with your Web site, you'll be able to see where customers came from to get to your Web site. You will be able to ascertain:

➲ Is the marketing mix within (both your traditional marketing plan and online marketing plan) driving visitors to your Web site as you planned? You can set the tactics up so that you can measure this.

➲ When visitors get to your Web site, where are they spending their time?

➲ How long are visitors staying on your Web site?

➲ What specifically are visitors reading and viewing?

➲ What aren't visitors reading and viewing?

Before you start to conduct research on the software vs. ASP (Active Server Pages) and Web analytics tools available, there are several factors to consider:

1. Determine the technical level of expertise for both the person responsible for pulling the correct data, and the person in charge of interpreting the data. Many companies incorrectly assume that a technical person can handle both. Make sure the person is in charge of interpreting the data is taking a marketing approach to what the data may reveal.

2. Real-time versus retroactive: ASP models tend to provide information in real-time, while log files take a retroactive approach to examining and making sense of web traffic. A large e-commerce Web site is an ideal candidate for an ASP-based solution. If your Web site doesn't change much, or you don't drive lots of traffic to your Web site via online marketing, software (usually a log file analyzer) is probably suitable for your needs.

3. Monthly visitors: do they fluctuate wildly? In general, the more
    visitors you have, the better off you'll be with an ASP-based solution.
4. What do you want to measure? Seems like an obvious question, but
    you'd be surprised by the number of people that don't know what they
    should be tracking. If you're one of them, think about what drives
    your business (both costs and revenue). You can track at least one of
    these major elements on your Web site.

Here's a short list of well known web analytics tool providers. They usually
offer either an ASP- or Software-based solution: ClickTracks, Coremetrics,
Omniture®, Unica® (formerly Sane Solutions), Urchin from Google, Visual
Sciences (formerly Websidestory) and WebTrends®, etc.

ANALYZING TRAFFIC

Here are some tips for analyzing the traffic results from web analytics
tools:

What to Track?
➲ As with any marketing initiative or plan, it's essential to establish a
   baseline from which you'll be able to measure all future activity. A
   web analytics tool will enable you to establish and define a baseline,
   for all the critical metrics that you determine to be most important to
   you and/or are associated with your marketing & sales objectives. It is
   likely that web analytics tool will have the capability to track a plethora
   of statistics—many stats which you may not need. Don't let the task
   of determining what to track overwhelm you. You only need to worry
   about tracking the pages and metrics that drive your business and that
   you need to measure your performance.
➲ Analyzing the data provided by web analytics tools can be very time
   consuming. Unless you plan to have a dedicated team member with
   ample time to dedicate to dissecting the results, you will want to
   keep the tracked metrics relatively simple—so as to ensure that you
   have time to turn the data into information each month. You know
   the old adage, data is just data until it is interpreted and becomes
   information—information from which you can make business
   decisions.

Who should Track?
➲ Analyzing the data provided by web analytics tools can be very time
   consuming. Unless you plan to have a dedicated team member with
   ample time to dedicate to dissecting the results, you will want to

keep the tracked metrics relatively simple—so as to ensure that you have time to turn the data into information each month. You know the old adage, data is just data until it's is interpreted and becomes information—information from which you can make business decisions.

When to Track?

➲ Make sure you track and capture all lead generation, online activity, and end results for at least three months before making any assessments. Capturing and analyzing 1-2 months worth of data won't take into account seasonality, or unusual external factors that may impact the traffic to your Web site. In other words, it's hard to draw any general conclusions about your Web site traffic based upon one point in time. Three months will allow you to establish a trend and allow you to have a more detailed view of the health of your Web site and its ability to meet visitor/customer needs.

➲ Establish a process for pulling and distributing a tracking report from within the analytics tool at least once a month. Make time to review, digest and interpret the results. If multiple people on your team receive the report, set a monthly meeting to sit down and review the results together. Don't just take the report and put it in a file.

How to use Tracked Information

➲ Whatever you do, first and foremost, stay consistent. Establish your baseline measurements, establish the activities that you want to track and don't deviate. If you change your reporting formats every month and keeping switching the activities you are measuring like flavors at an ice cream shop, it will be difficult to draw any conclusions or make any deductions.

➲ Be patient and stick to your plan. The key to properly utilizing the right online marketing tools is patience. For instance, it can take 12 to 18 months to optimize pages and email databases can take years to build up enough addresses for success. But do monitor the traffic reports to confirm that your other marketing tactics are performing and driving the traffic to your Web site.

➲ Monitor your regular or monthly reports for behavioral trends among visitors to your Web site. Are there areas of the Web site where traffic has fallen off over a 90-day period? If so, this should prompt you to ask the question why? What has happened within your business? Your industry? Among your competition? Maybe the content in that section is no longer valuable or relevant to your visitors or customers. If so, it's time to strategize on finding new content to replace the outdated content.

COMPLETION OF MODULE 4

Now you've completed the four phases of your Web plan. Refer to Module 5 for the outline and details of a properly constructed Web plan. The drive phrase is important to your organization whether you're promoting the new site or application to an internal audience (employees) or external (customer, partners, and prospects). We attempted to give you an overview of all the options and how to make the best choices. After completing modules 1-4, your new Web project is complete and has the proper strategies to be promoted to service all your target audiences successfully.

MODULE 5

PUTTING IT ALL TOGETHER

5.0 Putting it all together

Congratulations! You finished four of the five modules in this book. At this point you should have a good feel for all the various parts of a Web plan. Now your Web strategies will be defined and well organized. Also, you probably have many more questions than answers at this point. This is good – you're well on your way to succeeding on the new Internet. In the last module, we will tie all the steps together and even provide you with a basic Web plan outline.

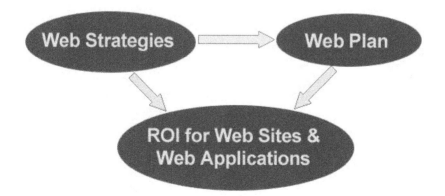

According to Forbes 500, most CEOs expect to generate almost 40 percent of their sales through the Internet within the next 10 years. It is more important than ever to have a Web plan. The traditional methods of marketing like television, newspapers, and magazines will continue to shift their attention more to the Internet and away from traditional media. Now is the time to understand how all of your Web strategies fit into a custom Web plan, enabling your firm to grow and succeed in the future.

5.1 ALL BUSINESSES ARE EBUSINESSES

Purpose: **Understand the new Internet and what it really means to be an ebusiness.**

Key Points: **In this section we will tie all the modules together, providing an overview of ebusiness and what successful implementation looks like. We provide a Web plan outline as a finishing step.**

According to Google, an ebusiness is defined as commerce conducted in cyberspace. It is the execution of real-time business processes with the assistance of Internet technologies. In essence, its business on the Internet -- but it also includes all the processes involved in operating a business electronically. The term 'ebusiness' is derived from such terms as 'e-mail' and 'ecommerce,' and includes not only buying and selling, but also servicing customers and collaborating with business partners.

Ebusiness is a buzzword used by many marketers in the new business environment. Many think of the term as selling products on a Web site where a transaction takes place. The reality is that every business is now an ebusiness.

Every company, organization, school, etc., does some type of business on the Web. If you are not thinking of your business as an ebusiness, you're probably falling behind. Customers expect you to do business on the Web and do it well. This ranges from providing simple contact tools to online account management. Today's business environment is not just about sales, marketing, product development, etc., but performing all the requirements of ebusiness. If you have not started to research and understand the true opportunities of ebusiness, you'd better get moving.

5.2 SUCCESSFUL IMPLEMENTATION

Purpose: **Understand what it is to successfully implement the right Web strategies.**

Key Points: **In this section we provide an overview of what a successful implementation looks like and the potential results that can be expected from it.**

As we mentioned in Module 1, if you've successfully implemented a new Web project, then consider yourself lucky. For the most part, new Web projects miss the target and end up being a waste of both money and time. Successful implementation is about creating and following a well-developed Web Plan.

As you go through your next project, observe the areas where the Web plan keeps you and your team on track and helps you hit your objectives. Furthermore, think of the feeling of relaxation and pride you'll have when the projects are completed and your firm grows and succeeds because of the plan.

Successfully building a Web project is about putting all the pieces together at all the right times. Indeed, just as one would implement a business plan, tactics change, competitive environments change, and new opportunities present themselves. The Web plan is simply your road map to success and is not set in concrete. We suggest that you perform analysis and updates on a regular basis (every three to six months is appropriate). With the plan as your tool, the road will be filled with fewer potholes.

5.3 DEVELOPING YOUR STAFF

Purpose: **Understand how to effectively use your internal staff either by themselves or in relation to an outsourced vendor.**

Key Points: **This section covers the process of improving the communication and skill level of your internal staff to complete your next Web project on time and within budget.**

During the past few years we have been asked more and more to assist with developing and implementing a plan for an internal Web team. This includes helping define job responsibilities, skill sets, improving skill levels, and formulating future Web plans. In fact, over the last few years the trend of taking Web development in-house has grown tremendously. The reasons for this trend include more skilled workers, reduced wages, and the importance of the Web in day-to-day business operations. When and how do you develop an internal staff?

KEY FACTORS

Knowing when to expand your internal staff and how to do it are difficult questions. You must take into consideration many factors including budget allowances, resources, and business objectives. You need to consider how the different skill sets of your internal staff will be utilized. If you do not have the right personnel, your projects may not get completed correctly or in a timely manner. On the other hand, if your personnel are under-utilized, the firm will lose productivity because people are sitting idle.

You will need to evaluate how important the Web applications are to your business objectives. Does it make economic sense to outsource them? Or, should you build a large staff that helps increase revenue and profit via online tools? The answers are easily found once you detail the financial implications and the specific objectives in your plan. Many firms understand the economic benefits of proper staffing numbers, utilizing an outsourced vendor only when the need occurs. If you run the numbers and it makes sense to expand your internal staff, how do you do it in the most efficient manner?

METHODOLOGY

Now that you've decided to expand your Web team, you must pay close attention to your methodology. First, evaluate the type of work/projects to be completed. Do they require personnel with front-end skills (graphic design, basic HTML, etc.) or strategy/consulting ability which could include writing content, setting up plans, understanding business processes, etc., or do you need staff that understands databases and complex, third party integration (java, .NNET, ASP, JSP, proprietary software, etc.). It is very difficult to find personnel with all three skill sets. In fact, the cost of having a person with all of these skill sets on your team may ultimately be cost prohibitive.

MISSING SKILL SETS

Once you've decided which type of personnel your objectives require, determine which skills are missing. By using this approach you can have an outsourced vendor "on call" for projects that come up infrequently. Identify the missing skills and connect them to the specific projects. For example, if you will build a customer service tool in the third quarter, create a list of deliverables and skills required. Let your outsourced partner know when and what the project will be ahead of time.

If you've chosen to use personnel with front-end skills, make sure they receive basic training on back-end applications and programming. They'll need to have some knowledge of the more complex functionality in order to work well with your outsourced vendor. They can also help with front-end recommendations based on their minimal knowledge of what is happening "behind the scenes" or in the code. In addition, if you've chosen personnel with programming or back-end skills, make sure they receive some front-end training. They should understand Photoshop files and usability issues. A little bit of training goes a long way when it comes to personnel who understand the complex issues.

Developing your internal staff is a key component of your Web plan. In the end, you'll need to understand when to add staff, what their job descriptions are, and how their skills help meet all the strategies of your Web plan.

WEB PLAN OUTLINE

At this point you've chosen the firm you're going to work with or identified your key internal players. In addition, you reviewed the first four modules and covered the first part of Module 5. Now you've reached the Web Plan. In the remaining pages of Module 5, we provide an outline for you to design and implement your unique plan. The plan is divided up into sections and provides a step-by-step process for your next project.

If you have an existing plan, this is your chance to make improvements. If you have no plan, time is running out on developing a successful Web strategy.

Get started today by calling 720.221.9214

XploreNet, Inc. www.xplorenet. com 720.221.9214

Visit www.xplorenet.com click on
to download the full web plan enter the code web20

WEB PLAN OUTLINE

MODULE 1: ANALYSIS:

MODULE 2: BLUE PRINT

MODULE 3: CONSTRUCT

MODULE 4: DRIVE

MODULE 5: TRACKING PROCEDURES

Visit www.xplorenet.com or call 720.221.9214 for more helpful ideas.

"Strong Mail Systems provides technology to get the message through. XploreNet provides the expertise and the understanding to bring your web presence alive. Together, the alliance brings together, for businesses, the coordination of their electronic strategies for web and email across marketing, sales, and support. The right message, to the right person, at the right time."

VP of Business Development
Simon Lonsdale

"My experience with XploreNet has been very good to say the least. All of the projects are bid very competitively, and project completion has always met or exceeded expectations in terms of timeline, quality, and professionalism. They are flexible and easy to work with."

Director Marketing Group
Raz Carcoana

"After three unsuccessful and frustrating experiences with web design and development companies, I stumbled onto a company called XploreNet. Any skepticism I might have had was quickly dispelled by their responsiveness, professionalism, and ability to deliver more than (what I thought) I needed."

Technical Director-Partner
Steve Johnson

"We were very impressed with what they were able to present to us and appreciated their patience."

CIPAC Coordinator
Roberta Smith, MSPH

"XploreNet Inc. has propelled our website into the future with new graphics, online donating capability, and user friendly, client manageable web information. The stylish ideas and clear presentations on each page bring the message about the Anchor Center to our families and donors in a simple but vibrant way."

Business Manager
Cynthia Wadle

"We operate a small business with a global market. XploreNet gives us the personal attention, range of web services, and support that we need as our business grows. In a market segment known for its abundance of technical expertise and lack of customer service, XploreNet stands out as a leader in providing customer focused services backed by deep technical knowledge."

Thomas Newman

"The City Club of Denver has been very impressed with XploreNet. With their very able assistance, we quickly transformed a tired-looking website into a much more vibrant and user friendly place to become acquainted with our organization. XploreNet's knowledgeable and extremely responsive professionals patiently steered us. I would recommend XploreNet without reservation."

President
Matt Hogan

"The Quizno's Corporation hired XploreNet to help develop a total rebuild of our existing Web site. Originally, the first phase of the project was scheduled for eight to ten weeks. Because of the Christmas holidays, and an unscheduled deadline, the work needed to be completed in only six weeks. XploreNet not only met the new deadline, but also delivered the final product with full functionality and no bugs."

Web Site Project Manager
David Dickenson

"I did an exhaustive search for top website developers to partner with Pure and found that XploreNet was far away the most attractive business partner. Most of this was due to Ian's close management of this relationship. He is very pro-active in identifying potential trouble spots and very careful that both of our companies are benefiting due to our partnership."

"I wanted to again congratulate you on the website – I think it looks terrific and you should be pleased. We are getting lots of compliments!"

General Manager
Molly Kenner

Acknowledgements

Bill would like to thank the following people for their help and support:
- My co-authors Dave Dixon and Michael Sevilla for their talents, partnership and friendship
- My wonderful wife and soul mate Karen for being a great example of humility, love, and strength
- Little Ella Baby for just being the most precious little thing I've ever seen
- Katie for being a great example of determination and drive
- Co founder Treg Meldrum for his faith and patience
- Michel Vallee and Gisela Stadelmann for their patience and support
- My sister Sally and parents Pete Young and Penny Asher
- My supportive Grandparents "The Gibbs"
- Rebecca Calhoun for her outstanding editing skills and phenomenal intelligence
- All my friends for understanding my obsessive compulsive entrepreneurial disorder

www.ingramcontent.com/pod-product-compliance
Lightning Source LLC
Chambersburg PA
CBHW051244050326
40689CB00007B/1053